City-HUBs

Sustainable and Efficient Urban
Transport Interchanges

City-HUBs
Sustainable and Efficient Urban Transport Interchanges

edited by

Andrés Monzón
Universidad Politecnica de Madrid, Spain

Floridea Di Ciommo
Technical University of Catalonia, Spain

CRC Press
Taylor & Francis Group
Boca Raton London New York

CRC Press is an imprint of the
Taylor & Francis Group, an **informa** business

CRC Press
Taylor & Francis Group
6000 Broken Sound Parkway NW, Suite 300
Boca Raton, FL 33487-2742

First issued in paperback 2018

© 2016 by Taylor & Francis Group, LLC
CRC Press is an imprint of Taylor & Francis Group, an Informa business

No claim to original U.S. Government works

ISBN-13: 978-1-4987-4084-5 (hbk)
ISBN-13: 978-0-367-13898-1 (pbk)

Visit the Taylor & Francis Web site at
http://www.taylorandfrancis.com

and the CRC Press Web site at
http://www.crcpress.com

Contents

List of Figures

List of Tables

Foreword

Cities play a key role in the European continent. Almost 60% of the total European population lives in urban areas, where two-thirds of Europe's gross domestic product is generated (European Union 2011). However, mobility patterns are clearly linked to urban density and the relative location of activities. Urban sprawl tendency in European cities has entailed difficulties for public transport to compete with the private car for many trips. Increasing motorized traffic – and its related negative impacts – can only be counterbalanced through efficient multimodal, integrated and intelligent public transport networks and services, as proposed in the 2011 Transport White Paper (European Commission 2011).

An efficient and reliable public transport system ensures this challenge is overcome. To this end, public transport networks should provide seamless door-to-door efficient services. However, the achievement of an integrated network necessitates adequate transfer facilities among modes. *Urban transport interchanges* provide an added-value link facilitating transfer among different modes. Enhancing multimodality ensures good services for future demand, avoiding over-dimensioning of the network while increasing its performance and attractiveness.

The European Commission launched several projects within the 7th Framework Programme aimed to improve multimodal interchanges. This book summarises the results of the *City-HUB project*, which assessed intermodal transport nodes in different European cities to identify good and bad practices. It develops an integrated model to design, operate and manage urban transport interchanges. This kind of initiative contributes to reducing the barriers to public transport use and therefore to the achievement of more sustainable and efficient urban mobility systems.

I want to thank the authors and the publishers for this book. It will facilitate researchers, practitioners and decision makers with a deep knowledge of the research done for improving existing projects and enhancing projects at new interchanges.

Patrick Mercier-Handisyde
European Commission
DG Research & Innovation
City-HUB Project Officer

Preface

This book has been developed based on the findings of the City-HUB project. The research project was funded under the European Union's 7th Framework Programme for Research and Technological Development. The project aimed to address, in a practical way, the research topic of "Innovative design and operation of new and upgraded efficient urban transport interchanges". The EU support made it possible to gather together practitioners, experts and researchers working together from the years 2012 to 2015.

The project followed a dynamic process involving the expertise of almost 100 stakeholders, information from surveys and data collected in 11 case study locations and lively discussions during 4 workshops held in 4 different countries. This book makes available the project's findings to a wider audience.

Consequently, all the contents of the following pages correspond to the contributions from a large number of people and institutions. Firstly, and most importantly, the real authors of this book are the partners of the project from nine different institutions and countries: Transport Research Centre (TRANSyT) of the Universidad Politécnica de Madrid (UPM), Spain; Közlekedéstudományi Intézet Nonprofit Kft, (KTI), Hungary; Transportøkonomisk Institutt (TOI), Norway; Centre for Research and Technology Hellas (CERTH), Greece; Panteia BV, the Netherlands; Transport Research Laboratory Limited (TRL), United Kingdom; VTT Technical Research Centre of Finland Ltd (VTT), Finland; Institut Français des Sciences et Technologies des Transports, de l'Aménagement et des Réseaux (IFSTTAR), France; and Centrum Dopravního Výzkumu v.v.i. (CDV), Czech Republic.

We include a list of the persons contributing to the project from each of the partner institutions (Table 0.1). We want to thank each of them for their hard and generous work during the 30 months of running the project.

The authors of each of the chapters of this book have done the necessary work of summarising in a logical way the large amount of information collected and produced by all the participants of the City-HUB project.

The editors wish to thank the numerous contributors. They particularly want to acknowledge the dedicated work of Derek Palmer for revising the

Table 0.1 City-HUB project consortium

Partner name	Country	Work team
Transport Research Centre (TRANSyT) Universidad Politécnica de Madrid (UPM) (Coordinator)	Spain	Prof. Andres Monzon Dr. Floridea Di Ciommo Sara Hernandez Dr. Rocio de Oña Nuria Sanchez
KTI Közlekedéstudományi Intézet Nonprofit Kft (KTI)	Hungary	Gábor Albert Dr. Imre Keserű Nóra Fejes Ádám Pusztai Álmos Virág Dr. Attila Vörös András Munkácsy
Transportøkonomisk Institutt (TOI)	Norway	Dr. Jardar Andersen Petter Christiansen Dr. Beate Elvebakk Julie Runde Krogstad
Centre for Research and Technology Hellas (CERTH)	Greece	Eftihia Nathanail Giannis Adamos Maria Tsami
Panteia BV	The Netherlands	Dr. Barry Ubbels Ricardo Poppeliers Arnoud Muizer Konstantina Laparidou Menno Menist
Transport Research Laboratory Limited (TRL)	United Kingdom	Derek Palmer Marcus Jones Heather Allen Clare Harmer Katie Millard
VTT Technical Research Centre of Finland Ltd (VTT)	Finland	Tuuli Järvi Juho Kostiainen Marko Nokkala Armi Vilkman
Institut Français des Sciences et Technologies des Transports, de l'Aménagement et des Réseaux (IFSTTAR)	France	Dr. Odile Heddebaut Lucia Mejia
Centrum Dopravního Výzkumu v.v.i. (CDV)	Czech Republic	Jan Spousta Jan Vasicek Michal Váňa

language and consistency of the draft chapters; also to Ana Barberan for coordinating the edition and to Divij Jhamb for giving us his design art for improving the content of the book with illustrations.

We hope that the final synthesis will offer readable and interesting information for researchers, transport planners, decision makers and all persons interested in achieving a sustainable, integrated, smart and efficient public transport system for better mobility.

Contributors

Giannis Adamos
Centre for Research and
Technology Hellas (CERTH)
Thessaloniki, Greece

Gábor Albert
Közlekedéstudományi Intézet
Nonprofit Kft (KTI)
Budapest, Hungary

Jardar Andersen
Transportøkonomisk Institutt
(TOI)
Oslo, Norway

Ana Barberan
Transport Research Centre
(TRANSyT)
Universidad Politécnica de
Madrid
Madrid, Spain

Petter Christiansen
Transportøkonomisk Institutt
(TOI)
Oslo, Norway

Floridea Di Ciommo
Transport Research Centre
(TRANSyT)
Universidad Politécnica de
Madrid
Madrid, Spain

Clare Harmer
Transport Research Laboratory
Limited (TRL)
Wokingham, United Kingdom

Odile Heddebaut
The French Institute of Science
and Technology for Transport,
Development and Networks
(IFSTTAR)
Villeneuve d'Ascq, France

Sara Hernandez
Transport Research Centre
(TRANSyT)
Universidad Politécnica de Madrid
Madrid, Spain

Tuuli Järvi
VTT Technical Research Centre of
Finland Ltd (VTT)
Espoo, Finland

Juho Kostiainen
VTT Technical Research Centre of
Finland Ltd (VTT)
Espoo, Finland

Katie Millard
Transport Research Laboratory
Limited (TRL)
Wokingham, United Kingdom

Andres Monzon
Transport Research Centre
    (TRANSyT)
Universidad Politécnica de
    Madrid
Madrid, Spain

Eftihia Nathanail
Centre for Research and
    Technology Hellas (CERTH)
Thessaloniki, Greece

Marko Nokkala
VTT Technical Research Centre of
    Finland Ltd (VTT)
Espoo, Finland

Derek Palmer
Transport Research Laboratory
    Limited (TRL)
Wokingham, United Kingdom

Ricardo J.M. Poppeliers
Panteia
Zoetermeer, The Netherlands

Ádám Pusztai
Közlekedéstudományi Intézet
    Nonprofit Kft (KTI)
Budapest, Hungary

Jan Spousta
Centrum Dopravního
    Výzkumu v.v.i. (CDV)
Prague, Czech Republic

Maria Tsami
Centre for Research and
    Technology Hellas (CERTH)
Thessaloniki, Greece

Álmos Virág
Közlekedéstudományi Intézet
    Nonprofit Kft (KTI)
Budapest, Hungary

Part I

Interchange concept

Chapter I

Introduction

*Floridea Di Ciommo, Andres Monzon
and Eftihia Nathanail*

CONTENTS

This City-HUBs book focuses on the design of 'sustainable and efficient interchanges', aiming to provide guidance and recommendations to enable seamless mobility, travel efficiency, user satisfaction and improved performance of the interchange. The main concern is to ensure that all people are given equal opportunities for reaching their destination, to optimise interconnections between alternative modes in the trip and, at the same time, ameliorate interchange space utilisation and integration in the city context.

The approach is based on a concise analysis conducted within the City-HUB project, which has the main objective of contributing to developing intermodality standards, minimum requirements, quality management tools, benchmark case studies and public transport service levels in Europe. All of these elements will provide the definition of operational good practices for urban interchanges and respective guidelines across European countries.

To this end, the content of this book has been organised into three parts. Part I is introductory, aiming to present the existing knowledge about interchanges and their role both as a public transport node and as a place in the city. After some insights into the City-HUB research project (Chapter 1), the book starts by presenting intermodality features in the context of complex and competitive mobility patterns in metropolitan areas (Chapter 2). It goes on to cover which aspects should be considered to select a typology of interchanges and presents a method for defining a typology for the interchange to be implemented in particular conditions (Chapter 3). Chapter 4 presents the rationale for promoting, developing and managing an intermodal interchange – the proposed City-HUB life cycle process.

Part II presents the findings from the analysis of different data collected during the City-HUB project: interviews with local stakeholders (questionnaire available in Appendix II), analysis of case study data collected through five extensive surveys to interchange users (questionnaire available in Appendix III) and the feedback from Transport Visioning Events in which various stakeholders have participated (participants listed in Appendix V). Within this section, Chapter 5 deals with the key aspects that should be taken into consideration when looking to improve the quality of an intermodal interchange. Chapter 6 presents how to acquire information about the needs and requirements of users and identify priority action domains for obtaining an attractive urban interchange from the users' point of view. Finally, Chapter 7 presents the City-HUB model summarising all the elements to consider when developing an urban interchange. Appendix IV complements this section by applying some of the ideas presented to three case studies.

Part III presents the benchmarking exercise, conducted through the analysis of good practices from 11 case studies of urban interchange across Europe. Chapter 8 analyses the performance of selected interchanges in different domains, such as business models, services and facilities, governance and the coordination of transport demand. This chapter is completed with some factsheets that list the main features of each analysed interchange (Appendix I). Finally, Chapter 9 presents the overall recommendations for developing urban interchanges.

1.1 INTEGRATED TRANSPORTATION IN AN ERA OF CHANGE

Over recent decades, integrated transport has become one of the most prominent topics and areas of interest, in which the contribution of the EU has been realised by the formation of concrete policies, the creation of transport funding facilities and the definition of priorities for different transport networks (Adamos et al. 2012).

As a follow-up to the 2011 White Paper (European Commission 2011), in 2013, the European Commission produced the 'Urban Mobility Package', which settled the concept of 'Sustainable Urban Mobility Plans' (SUMPs) (European Commission 2013a). The concept defines the basic characteristics that a modern and sustainable mobility and transport plan should be comprised of, and includes *the balanced and integrated development of all modes* as one of them.

Although the European Commission has been addressing the upgrading of urban interchanges in order to increase public transport use (European Commission 2001), the results in terms of impacts on increasing intermodal use remain very low. It is therefore essential to build an understanding of the key factors of interchanges that affect public transport and intermodal/multimodal use.

Good practices for urban passenger interchanges in Europe were first presented in the Group for Urban Interchanges Development and Evaluation

(GUIDE) project (Terzis and Last 2000). In their analysis, interchanges were considered as an inescapable feature of supporting seamless public transport.

Experience has shown that the introduction of interchanges increases the use of public transport (Monzon et al. 2013; Di Ciommo et al. 2009; Brons et al. 2009). The European Commission has started on a path towards the upgrading of urban interchanges for increasing public transport use along with many public transport authorities worldwide (European Commission 2001). A key factor for increasing the use of public transport and soft modes, and the reduction of private motorised trips, is related to users' perceptions and preferences in respect to time savings and time use during the intermodal trip (Crozet and Joly 2004). Various other studies using a pure utility approach show that transfer time is perceived as negative and as a disutility (Mackie et al. 2001). Paying more attention to the physical location where the transfer happens and increasing comfort, safety and security perception during public transport use will be key factors for attracting additional users of intermodal trips. This is a way to reduce car dependency in sprawling cities: the trade-off is no longer between planning more to travel less, but between planning more to travel better (i.e. low carbon transport). Therefore, the EU Directorate of Transport Research identified urban interchanges as key infrastructures for sustainable mobility, and defined a proposal for extensively analysing urban European interchanges.

The novelty of the City-HUB approach has been to focus attention on the physical space where people interchange between two modes of transport. It has adopted a holistic approach, including three key domains for managing, operating and using an interchange: governance, service and user requirements.

1.2 CITY-HUB VISION OF INTERCHANGES

The following chapters will focus on defining different issues of multimodal trips in urban areas. However, although European funding for intermodal policy and technology projects is 10% of the total transport research budget (TRI-Value 2014), the results in terms of impacts on increasing intermodal trips remain very low. This means that cities require greater effort to improve the understanding of key factors for increasing public transport trips through improving intermodality, and that this process of transferring from one mode to another usually takes place in more efficient conditions at modal interchanges. Different multimodal trips require different types of interchanges, as presented in Chapter 3.

The City-HUB project starts from a holistic approach, taking into account these different perspectives and including elements affecting the quality of an interchange for the transport services, the different stakeholders and the city itself (see Figure 1.1).

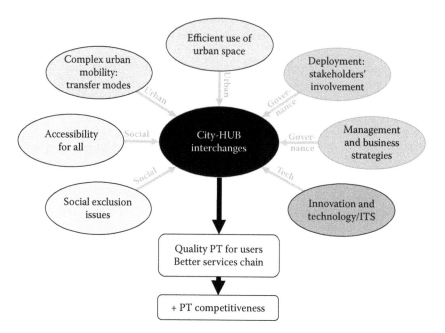

Figure 1.1 City-HUB vision of interchanges.

The City-HUB vision of interchanges starts by establishing the needs of urban mobility patterns and examining how to use scarce space in urban areas for transport interchanges. The function of an interchange station is to reduce the distance between two different urban areas and, therefore, to facilitate multi-activity patterns. The location and design must pay attention to social aspects such as accessibility and social exclusion issues. Information and communication technologies have a key role in assuring solutions and their efficiency. However, all these technical and urban mobility aspects should fit within adequate business models for strategic stakeholders to promote and manage interchanges.

The main added value of the City-HUB vision is that it provides a multidisciplinary approach, which amalgamates relevant scientific and policy aspects. These refer not only to specific mobility issues, but also to technology, economy, land use planning and social concerns. Figure 1.1 outlines the main priorities of the City-HUB interchanges.

One of the main priorities of the City-HUB vision is related to the current concerns of EU policymakers regarding urban mobility, in particular, the need to *improve the quality of public transportation services*. According to the Transport White Paper (European Commission 2011), mobility is vital for the internal market and for the quality of life of citizens as they enjoy their freedom to travel. In this context, the quality, accessibility and reliability of transport services will gain increasing importance in the coming years, inter alia, due to aged population growth, urban sprawl and the need to promote public transport (Ewing et al. 2008; Vuchic 2005). Comfort, easy access, reliability, attractive frequencies of services and intermodal integration are the main characteristics of service quality. The availability of information regarding travelling time and patterns alternatives is equally relevant to ensure seamless door-to-door mobility.

Many metropolitan authorities are implementing policies designed to promote public transportation through increasing the investment in new infrastructure and improving the quality of the public transport services offered. However, in spite of the advantages that promoting public transportation has in terms of the reduction of externalities (pollution, carbon emissions, noise, congestion and so on), investing in new infrastructure is often very burdensome for municipal and regional governments, who presently are having to face serious budgetary constraints (Di Ciommo et al. 2009).

In this context, *urban transport interchanges* play a key role as components of public transport networks in *facilitating the links between different public transportation modes*, particularly the connection of bus services to the subway and metropolitan railway system (Vuchic 2005). Research literature shows that the benefits of urban interchanges mainly relate to time saving, better use of waiting time, urban integration and improving operational business models (Di Ciommo 2002).

In summary, the City-HUB project has developed an integrated model which embraces the different aspects of an interchange in order to decrease the barriers to the use of public transport, to improve the quality and to propose a business model related to the interchange typology. The approach will help frame pathways towards obtaining maximum efficiency by upgrading existing urban interchanges or by building new ones and making these more efficient and accessible to all users.

1.3 ROAD MAP TOWARDS THE CITY-HUB MODEL

This section proposes the road map for the City-HUB model. It aims to provide guidelines to support stakeholders in realising successful interchanges based on the results of the City-HUB project related to the integration of design and management of an interchange in response to travellers' desires. The City-HUB project has based its research activities on detailed knowledge on transport intermodal systems, the consultation process with stakeholders and users' experiences and expectations. The implementation process was carried out in different situations and scenarios across Europe using selected case studies based on new and improved urban interchanges.

Within the City-HUB project, the consultation process was designed to understand the key factors for efficient interchanges from the point of view

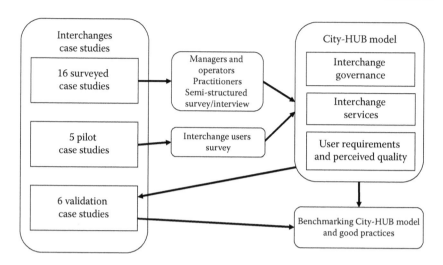

Figure 1.2 Process to develop the City-HUB model.

of stakeholders and users. After a comprehensive literature review, this process was based on analysing operations and performance data collection and surveys. Figure 1.2 shows the process based on the analysis of 21 selected interchanges. The lessons learnt from these case studies served as inputs for validating the City-HUB model through six additional case studies.

The semi-structured interviews of operators and managers in 16 surveyed case studies (see Appendix II) set the basis for developing the analysis and proposals for the governance and services of interchanges. The 2,000 attitudinal surveys in the five pilot case studies (see Appendix III) served to identify the key factors for travellers at interchanges. This included their perceived quality of the existing services and the need for improvement. In summary, this process allows us to define the City-HUB model that considers all the aspects for interchange deployment and management and also its integration in the local business and urban fabric. This model corresponds to the multifaceted vision of stakeholders and users.

Eckstein (1975) emphasised that the selection of case studies could provide maximum analytical leverage. A least likely and most likely approach can thereby make it possible to find robust support for theories and hypotheses. A least likely approach selects cases which are at the limit of the theory's boundaries, while a most likely approach could identify good reasons for refusing a theory since it selects cases from the heart of the theoretically defined scope. In order to validate the City-HUB model, it is necessary to emphasise the careful selection of the case studies (Lijphart 1971). Our aim in dealing with the various case studies has been to select cases that are comparable for specific elements, but which are also diverse (Ragin et al. 1996). The 27 selected case studies are listed in Table 1.1, which includes

Table 1.1 Interchanges analysed in City-HUB project

Surveyed interchanges and location	Transport services				Road modes	Rail modes	Daily passengers
	Local	Regional	National	International			
Leppävaara Station, Espoo, Finland	X	X	X		X	X	25,500[a]
Lillestrøm Station and Bus Terminal, Norway	X	X	X		X	X	11,000
Bekkestua Interchange, Bærum, Norway	X	X			X	X	4,500
Plaza Castilla Interchange, Madrid, Spain	X	X			X	X	199,500[b]
Méndez Álvaro Coach Station, Madrid, Spain	X	X	X	X	X	X	180,000[c]
New Street Station, Birmingham, United Kingdom	X	X	X		X	X	140,000
Reading Station, Reading, United Kingdom	X	X	X		X	X	52,000
Kings Cross-St. Pancras Station, London, United Kingdom	X	X	X	X	X	X	80,000[d]
Leiden Train Station, Netherlands	X	X	X		X	X	60,000
Den Bosch Train Station, Netherlands	X	X	X	X	X	X	59,000
Érd Intermodal Terminal, Hungary	X	X	X		X	X	21,000[e]
Gares Lille Europe & Lille Flandres, France	X	X	X	X	X	X	118,500[f]
Intercity coaches of Magnesia Interchange, Volos, Greece	X	X	X		X		2,500
Macedonia Coach Terminal, Thessaloniki, Greece	X	X	X	X	X		22,500[g]
KTEL Kifisou, Athens, Greece	X	X	X		X		26,000[h]
Prague Terminus Dejvická, Czech Republic	X	X	X		X	X	150,000[i]
(5) Pilot case studies							
Moncloa, Madrid, Spain	X	X	X		X	X	287,000[j]
Kamppi, Helsinki, Finland	X	X	X	X	X	X	57,000[k]
Ilford Railway Station, London, United Kingdom	X	X			X	X	21,000[l]

(Continued)

Table 1.1 (Continued) Interchanges analysed in City-HUB project

Surveyed interchanges and location	Transport services				Road modes	Rail modes	Daily passengers
	Local	Regional	National	International			
Köbánya-Kispest, Budapest, Hungary	X	X	X	X	X	X	155,500[m]
Railway Station of Thessaloniki, Greece	X	X	X		X	X	166,500[n]
(6) Validation case studies							
Gares Lille Europe & Lille Flandres, France	X	X	X	X	X	X	118,500[f]
Utrecht Centraal, Utrecht, Netherlands	X	X	X		X	X	285,000
Oslo Bus Terminal Vaterland, Norway	X	X	X	X	X	X	27,500
Paseo de Gracia, Barcelona, Spain	X	X			X	X	100,000
Prague Terminus Dejvická, Czech Republic (future development)	X	X	X		X	X	N/A[o]
Intermodal terminal of Miskolc, Hungary (future development)	X	X			X	X	N/A[p]

a   Train + buses.
b   Buses + metro.
c   Buses + metro + coaches.
d   At three peak hours.
e   Metro + buses + tram.
f   At Lille Flandres, 70,000 take the train + 40,000 crossing (using services or going into the shops); and 8,500 train-only passengers at Lille Europe.
g   20,000–25,000 passengers per day.
h   25,000–27,000 passengers per day.
i   Metro station is used by around 120,000 passengers.
j   Metro + buses + coaches.
k   Metro + buses.
l   Rail-only.
m   Train + buses + metro.
n   Buses + train.
o   150,000 at the existing interchange.
p   5,000 at the existing station.

their location (see also Figure 1.3) and main characteristics. Data collection for the case studies was carried out throughout the City-HUB project lifetime (i.e. 2012–2015).

The final set of six validation case studies was used to assess good and bad practices and to identify improvement potential for developing a successful interchange. From some of the validation case studies, it could also be understood that the model was assumed to be more useful for larger than for smaller interchanges. For instance, the more modes and stakeholders involved, the stronger the need to have a holistic model such as City-HUB. Likewise, the more passengers, the greater the need to separate passenger flows, offer relevant services and so on.

When moving through the sections, the reader will first be informed about the significance of interchanges in the domain of intermodal transportation,

Figure 1.3 Geographical distribution of case studies.

European policy and goals for increasing interchange effectiveness and the assessment of the main components of an interchange. Throughout the first two sections, a synthesis of the literature and the findings of the City-HUB project depicting the necessary actions and steps to be undertaken towards delivering a successful interchange are presented, in relation to:

- Governance, which incorporates the identification of the stakeholders and interchange users and their roles, methods for developing a cooperative scheme for efficient and mutually accepted decision-making, the development of business models and monitoring and assessment of the implementation performance.
- Services, which are related to the physical design, transportation modes and information provision at the interchange about the interchange and the trip, and visitors' facilitation during their stay at the station.
- User needs and expectations regarding the interchange design and operation, which involves conducting surveys for data collection about expectations and perceptions of service quality assessment.

Implementing the good practice guidelines is the main topic of Part III. In this last section, the guidelines developed in Part II are applied in selected interchanges, validating them and providing valuable and practical feedback on their actual implementation.

Chapters 2 through 6 analyse the key elements for organising and managing an interchange from the operators', users' and stakeholders' points of view, while Chapter 7 describes the City-HUB model and Chapter 8 presents the cross-case analysis using the validation process of the City-HUB model.

Chapter 2

Why interchanges?

Understanding intermodality

Eftihia Nathanail, Giannis Adamos and
Maria Tsami

CONTENTS

2.1 ROLE OF URBAN INTERCHANGES IN PASSENGER INTERMODAL MOBILITY

Although transport and mobility are two different notions, they are strongly interrelated. Transport includes infrastructure and services and works as the means to achieve mobility, while, on the other hand, mobility is the behavioural outcome of people's need to travel for achieving every-day activities or for consuming products, goods and services (European Commission 2007; Geurs et al. 2009; Levinson 2010; Wang et al. 2015).

Where is the balance between what transport can provide and how the desired mobility level is achieved? How can this balance be made compatible with a low-carbon society? With a dependency on mechanised trips, the pursuit of sustainable and inclusive mobility is a challenge (Di Ciommo and Lucas 2014). The solution is widely perceived to lie in an integrated approach, addressing and combining joint initiatives and integrated policies in the sectors of transport, environment, spatial planning, land use, new technologies and economy. In an era of rapid change, both citizens and decision makers need to address contradictory mobility issues. When technological instruments for improving daily lives exist, the economic crisis restricts the potential use of these instruments (e.g. implementation of real-time information screens).

Mobility appears as a fundamental component of people's daily lives. The European Commission's White Paper on Transport Policy of 2001 pointed out that intermodality is a key factor of daily mobility and identified it as the means for ensuring seamless travel at the metropolitan and urban level in Europe. Intermodality continues to appear in subsequent documents, with the Mid-Term Review of the Transport White Paper in 2006 (European Commission 2006), the Action Plan for the Deployment of Intelligent Transport Systems in 2008 (European Commission 2008) and the EU Transport Policy in 2011, in which three pillars are considered: people, integration and technology.

In a more recent document, Thematic Research Summary: Passenger Transport (European Commission 2013b), the focus is placed on integrated transportation services through information provision and intermodal coordination, where different transportation modes are interconnected (Figure 2.1).

Intermodal mobility necessitates the concurrent adaptation of the following elements:

1. The geographical location.
2. Interconnected transportation modes.
3. Infrastructure (physical) and technological facilities.
4. The system's operation, including the cooperative schemes required for the interconnection of the transportation modes, operators and system providers.

Although national governments and local authorities make efforts to persuade travellers to switch mode, it seems that public transport still is

Figure 2.1 Liverpool Interchange, London, United Kingdom. (Courtesy of Jan Spousta.)

not capable of competing with the private car for many trips (Grotenhuis et al. 2007; Graham-Rowe et al. 2011). From the customer's perspective, the quality of services provided by public transport is not at a satisfactory level that would urge them to replace their car with other modes, that is, bus, train, tram or a combination of these. But, as the world becomes more urbanised, there is a strong need for urban public transport to provide a viable alternative to individual car transport. However, the availability of opportunities for direct journeys when using public transport is limited. For this reason, the majority of trips require interchange zones, and hence areas which encompass one or more interchange facilities and public spaces used for access and/or transfer (Liu et al. 1997).

Interest in the quality of urban passenger interchanges in Europe has been growing since the beginning of this century and this has resulted in several research studies. The GUIDE project (Terzis and Last 2000) was probably one of the first studies to identify existing European research and practices concerning urban interchanges. It highlighted that an interchange is an inescapable feature of public transport and assessed best practice in terms of functional specifications and design. The project defined distinct aspects of interchanges that were found to be of relevance for seamless public transport systems (such as accessibility, facilities, image and information provision).

Urban transport interchanges mainly interconnect different transportation modes, which complement each other to accommodate a person's

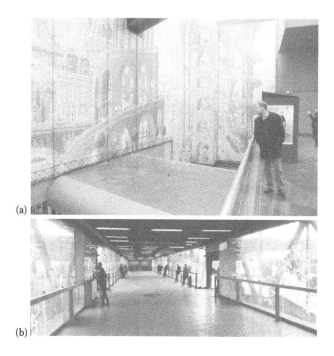

(a)

(b)

Figure 2.2 Interchange attractiveness: Lille Europe, France (a) vs. Köbánya-Kispest, Budapest, Hungary (b). (Courtesy of Jan Vasicek (a) and Andres Garcia (b).)

journey from its origin to its destination. When interconnection is properly designed and managed, many benefits arise for users, including time-saving owing to reduced transfer time and more efficient travelling. This can be achieved through the coordination of public transport services and the provision of integrated information to the users. Another aspect of interchanges is the provision of services to travellers waiting to change modes, thus utilising the facility for purposes other than travelling. Of course, with this transformation, interchanges become attractors for visitors, which increases the usability of the facilities, improves the image of the urban area and can promote the development of local businesses (Figure 2.2).

On the other hand, to ensure integrated and efficient transport of passengers between different transportation modes and between several routes, an urban transport interchange should (Pitsiava-Latinopoulou and Iordanopoulos 2012):

1. Provide reliable and adequate level of the direct services offered, such as information and ticketing.
2. Develop satisfactory facilities serving the transfer in service areas and waiting areas/platforms, through offering amenities, Internet access, comfort, and so on.

3. Provide adequate accessibility to the site for all users (especially the disabled).
4. Afford assistance to travellers with navigating aids, so that they can find their way from where they are to where they wish to go, both within the interchange, as well as to and from the local vicinity (way-finding).
5. Offer easy and seamless navigation and movement of users, improving also their understanding, enjoyment and experience (legibility).
6. Allow users to move around the interchange under several alternatives, providing at the same time clear connections to existing routes, facilities and services (permeability).

The sustainability aspects of transport interchanges usually focus on environmentally friendly services and infrastructure. Sintropher (2012) includes sustainability among the three main criteria for evaluating urban interchanges. Apart from sustainability, integration and technical design are the main components to be analysed. Edwards (2011) considers transport interchange design as an urban realm component (Figure 2.3). According to his view, social sustainability is related to sustainable modes of transport. It is generally accepted that an appropriate interchange design has to consider all modes of transport, especially soft modes (i.e. cycling and walking) (Taylor and Mahmassani 1996; Tsami et al. 2013a,b). The PRESTO project (Dufour 2010) discussed the appropriate interchange infrastructure and operation to accommodate cycling needs and promote soft modes of transport. Investment in low-cost forms of transport was studied by Naude et al. (2005) in an interchange design with a focus on pedestrian needs.

Regarding accessibility, transport interchanges face a challenge to be equally accessible to all kinds of transit users. Two kinds of barriers occur, especially regarding accessibility for people with disabilities. The first is related to information provision and the second is related to the physical

(a)   (b)

Figure 2.3 Transport Interchange as an urban development component: Stratford Interchange, London, United Kingdom (a) and Lille Flandres, France (b). (Courtesy of Jan Spousta (a) and Jan Vasicek (b).)

movement of interchange users. In the first case, the unpredictability of the journey experience is indicated as a significant problem (TfL 2009), while the physical problems are mostly related to boarding and alighting public transport modes at interchanges, as well as crowding, which restricts smooth movement. Accessibility is affected by many factors, such as mobility, quality and affordability of travel options, connectivity of the transport system, mobility substitutes and land use features. Walking and public transport seem to be the most important modes for people with disabilities (Litman 2012).

The quality and location of a transport terminal, as well as the connection between links and modes, also affect the accessibility of the terminal (Litman 2012). The need for universal design in transport interchanges is associated with the need for providing efficient movement for all categories of travellers. Universal design requires disability access and the level of service in this case is the degree to which transport facilities and services accommodate people with reduced mobility.

Apart from sustainability and accessibility aspects, travellers tend to use an interchange based on their personal travel needs and perceptions. User perception is most commonly considered when assessing an interchange (Litman 2012) and thus it is crucial to understand users' opinions. It is commonly accepted that the design and operation of a transport interchange may influence the physical experiences and psychological reactions of a traveller, and thus an efficient design and operation should attract travellers and be linked to the sustainability of an interchange. The GUIDE project (2000) underlined the major influence that interchange design could have on the general perception of travellers using public transport.*

2.2 BARRIERS AND RECOMMENDATIONS FOR INTERMODAL MOBILITY

Based on the results and findings of recent European projects on intermodality (e.g. GUIDE, MIMIC, LINK, ORIGAMI, CLOSER, INTERCONNECT, City-HUB itself), a number of 'barriers' to intermodal mobility emerge. Despite the efforts made at EU level in forming policies to enhance passengers' intermodal mobility, there is still not sufficient in-depth elaboration of them in transportation operations.

Data on intermodal mobility are generally nonexistent or insufficient and there is poor knowledge of the market segments for intermodal transport.

---

* Part of the content of this chapter has been developed previously in the Deliverable D2.1 of the City-HUB project, 'Review of theory, policy and practice', by Barry Ubbels, Derek Palmer, Konstantina Laparidou, Ricardo Poppeliers, Menno Menist, Marcus Jones, Katie Millard and Clare Harmer.

Insights on modal split and market potential for intermodal mobility are usually addressed in terms of national travel surveys, but these are not sufficiently focused on intermodal mobility (Collet and Kuhnimhof 2008). Also, there is heterogeneity of *intermodal knowledge* among European countries and a lack of relevant common data available, which particularly impedes the planning of seamless long-distance mobility (Riley et al. 2010).

Another barrier is the existing legislation across Europe regarding intermodal mobility, which is still incomplete and rather rudimentary; while it does not prevent *intermodal mobility*, it does not promote it either. A key bottleneck is the *modal* approach of most existing legislation, which is not conducive to modal integration in an urban or interurban context. However, as was noted in a recent presentation (EITR 2012), prescribing details for a pan-European legal framework in this respect will not serve the vision of having an intermodal and integrated mobility system, as this has to be found and decided at a local level.

Proper charging for the use of transport infrastructure is one of the unresolved issues that affect intermodality. The future integrated mobility systems will have to include some form of the *polluter pays principle* in their costing and ticketing structures so that while they make it easy to travel seamlessly they do it in a sustainable way. Some studies suggest that improvements to public transport and reducing the cost of using it could alleviate people's reluctance to accept road pricing (EITR 2012).

The CLOSER project formulated a list of the main interchange-related barriers and the recommended remedial actions, depicted in Table 2.1 (Auvinen et al. 2012).

Table 2.1 Terminal-related barriers and remedial measures

Barriers	Recommendations
Involvement of several authorities in the decision-making processes.	Definition of a procedural framework with explicit definitions of roles of each stakeholder. Distinction of ownership and operational responsibilities.
Conflicts of economic, societal or environmental interests of stakeholders.	Integration of transport planning and land use decisions. Definition of assessment criteria and consultation procedures for interchange implementation.
Shortcomings and gaps in the legal framework to promote comprehensive intermodal mobility systems.	Breakdown of EU-level transport policies to the practical level, e.g. strategic spatial distribution and mode choices of interchanges. Explore the need for a harmonised European regulatory frame for interconnections.
Insufficient public resources to finance terminal development projects.	Targeting public sector funding schemes and instruments that facilitate private sector involvement.
Absence of a master plan for interchange terminals.	Regular updates of 'master plans' of interchanges.

2.3 CHALLENGES OF EFFICIENT INTERMODAL MOBILITY

As stakeholders appear to be the key players in the planning and operation of interchanges, supporting their engagement and enhancing their cooperation in the planning, design and operation of urban interchanges is the main challenge that has to be addressed.

There are three groups of stakeholders that affect intermodal mobility: decision makers, service providers and end users. The challenge of efficient intermodal mobility is oriented to satisfy each specific group and its needs and requirements. The following sections present the challenges for each group of stakeholders.

2.3.1 Policy and governance

Existing transport policy and the involvement of governments are the two principal factors in interchange design and development. Multiple levels of government may be involved depending on the scope of the interchange. Their roles may not always be clearly defined, and financing can be the overarching problem for any further actions towards interchange development. Some findings about good practices are presented here.

2.3.1.1 Policy management and coordination

Policy consistency and clear institutional structures in operations and management are two of those key elements, as the project Towards Passenger Intermodality in the EU (European Commission 2004) has identified.

An issue in the Netherlands, for instance, is where the management of public transport is a local or regional responsibility. The national government is involved in the transport between the nodes, but usually nodes operate under the jurisdiction of local authorities, unless they constitute interchanges of national importance (e.g. Rotterdam Port and Schiphol Airport and seven major station projects). As it is not always clear which type of stakeholder takes the lead, setting in advance a visionary lead seems to be a key precondition for successful interchange development and management.

Inter- and intra-organisational networks can build trust and mutual understanding amongst stakeholders. It might be necessary to set mandatory requirements and provide guidelines to ensure systematic cooperation between the stakeholders, through a coordination plan which can allocate responsibilities and solve management problems.

2.3.1.2 Financing of interchanges

It appears that partnerships can be successful, but that good preparation and communication is important. Public–Private Partnerships (PPPs) can offer a very useful tool for realising public transit–related urban development, which the public bodies might not realise on their own. However, it should also be kept in mind that they do have potential disadvantages and the forging of such partnerships should thus be carefully considered on a case by case basis.

2.3.1.3 Sustainability of interchanges

Sustainability is an important policy aim in contemporary society, with the concept being included in design and evaluation frameworks for urban interchanges (see e.g. Network Rail 2011 and TfL 2009). Sintropher (2012) focused on three criteria for evaluating urban interchanges: sustainability, integration and technical design. Interchanges should deliver value for money and provide a positive economic, social and environmental impact.

Transport interchanges play a significant role in the whole network along with urban development. The emerging issue of developing sustainable transport interchanges is part of the interchange design objectives, and soft modes of transport are considered a crucial part of this target. Social sustainability is related to sustainable modes of transport. Soft modes of transport, such as cycling and walking, are expected to reduce traffic congestion and accidents and improve urban mobility and safety.

The impact of the construction or redevelopment of an interchange can have consequences on land uses and disrupt visual settings as the operation stage may cause noise, vibration and air pollution. The impact of design on the environment should be included in cost–benefit analyses, but are difficult to monetise in many cases.

Sustainable design may be costly, and in the end it is a trade-off between the mobility, safety, environmental and economic priorities that are set for a transport interchange project.

2.3.1.4 Security at interchanges

Urban transport interchanges are public facilities, visited daily by millions of passengers, which makes them easy and attractive targets for terrorist attacks. The threat of terrorism and the urban transport security issue gained tragic recognition in EU policy after the Madrid bombings on 11 March 2004. The European Council's Declaration, adopted two weeks later, acknowledged the fight against terrorism and the full implementation of the European Security Strategy of 2003 as a matter of urgency and called for a long-term strategy to address all the factors which contribute to terrorism. With regard to specific measures to protect transport, the European Council called for 'strengthening of the security of all forms of transport systems, including through the enhancement of the legal framework and the improvement of prevention mechanisms'. This includes work to develop further EU transport security standards in coordination with relevant international organisations and countries.

2.3.2 Service providers

2.3.2.1 Transport interchange operations

Intermodality aims to create efficient and seamless connections between transport modes by engaging local transport operators, coordinating access at the interchange, and providing integrated scheduling, ticketing and information. This was indicated as one of the most important factors for attractive transport in the SPUTNIC project (2009).

The HERMES (2011) project highlighted that the lack of cooperation among operators was mainly due to different ticketing systems, the lack of incentives for operators and tariff systems that were not integrated. Coordination among public authorities, and between the public and private sectors, has been identified as one of the overarching issues by transport operators and interchange managers. Specifically, managers reported that in 60% of cases, cooperation failed due to the lack of coordination between operators, whereas 25% failed because of different ticketing and tariff systems. In addition, merely one-third of the (investigated) operators had agreements for information services and timetable synchronisation.

2.3.2.2 Management and maintenance

The lack of adequate funding for maintenance leads to the deterioration of interchanges and might ultimately reduce their attractiveness for passengers. Studies of interchange users have concluded that general appearance is an important feature. In addition, lack of maintenance might lead to higher costs when upgrading, to catch up with previously neglected measures.

The management of interchanges involves, amongst other issues, the establishment of the rights and responsibilities of stakeholders and the facilitation of station maintenance and cleaning (Network Rail 2011). An effective organisational framework for operations and planning is a priority for infrastructure managers as station facilities are shared amongst different actors. For example, each mode of transport manages its own space, while the commercial activities (shops, restaurants, etc.) are concessions to intermediaries. Priority topics for interchange management are the engagement of local stakeholders to ensure coordinated maintenance and other responsibilities for public space and access routes, the formulation of standards for interchange activities, and the effective management of deliveries. Furthermore, interchange managers should maintain the minimum hygiene standards and facilitate to the best possible extent the transition of passengers.

2.3.2.3 Safety and security

The threat of terrorism and urban violence in public transport has also recently become an important issue for operators. Whilst efforts undertaken by public transport companies across the EU and the world have been irregular, many operators have in recent years increased their safety and security arrangements, reacting to the potential threat from terrorist attacks. Although every urban transport network is unique in its size, complexity and the threats it faces, common elements of security efforts undertaken are still observable.

2.3.2.4 Revenue generation

Well-designed and accessible interchanges that accommodate many trips on a daily basis are attractive for many businesses, such as transport operators, retailers, services and other commercial and public activities. Station owners and managers can generate revenues from renting space. Revenue generation from interchange activities could lead to a significant funding stream to support maintenance and operation costs.

2.3.3 Interchange users

Passengers' decision-making processes are complex and influenced by their own personal characteristics and needs. It is important to understand them in order to be able to provide better services and prioritise actions for improvement. Interchanges are the 'principal shop window' for the public transport system and, as such, the impression they give will be a major influence on perceptions of public transport (Terzis and Last 2000).

Understanding passengers' points of view is essential for the efficient design and management of interchanges (Monzon et al. 2013). Passengers' requirements depend on their mobility needs and trip purpose. There are also differences in needs between users and nonusers (potential users) of interchanges, and men and women (Pirate 2001). In general, the expected benefit can be expressed in time-savings and convenience, and comfortable and safe travelling. At interchanges, this relates to efficient and safe changes between modes and the availability of services that facilitate seamless travel (bicycle stands, parking lots, bus lanes, etc.).

When passing through an interchange, travellers usually need to wait and spend time in that location. Commuters generally want to travel as quickly as possible, whereas leisure travellers are less resistant to long waiting times at interchanges. When other services besides the transport function itself are offered, or shops are located at the interchange, the quality of the time spent waiting may improve.

Additional factors affecting the design and management of an interchange are the special requirements of those travellers with mobility impairments. Improving accessibility for people with reduced mobility is an important urban transport policy objective in the European Policy for Urban Mobility (see for instance Actions 3 and 5 of the Action Plan on Urban Mobility of the EC [European Commission 2009] and the main issue of service

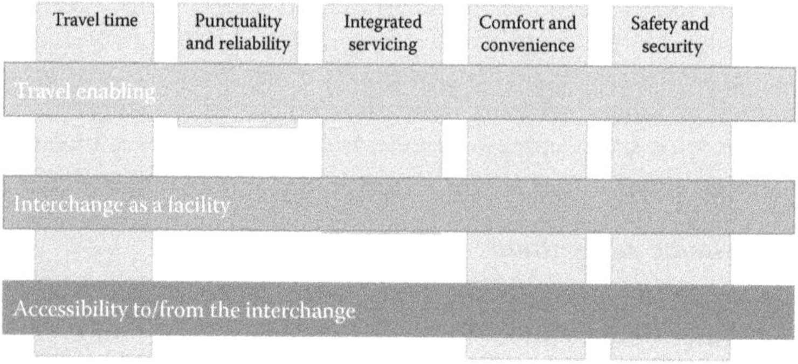

Travel time	Punctuality and reliability	Integrated servicing	Comfort and convenience	Safety and security

Travel enabling

Interchange as a facility

Accessibility to/from the interchange

Figure 2.4 Interchange quality parameters.

quality and passengers' rights in the White Paper on Transport [European Commission 2011]). Accessibility needs depend on the type of disability a user may experience. Interchange accessibility and social exclusion aspects are important and, as such, have been identified as an important element within the City-HUB project.

There are three main categories which aggregate the expectations of the users of an interchange and are related to interchange accessibility and offered interchange facilities (Figure 2.4).

An interchange manager's objective is to achieve passenger satisfaction and loyalty through the provision of a better quality of service, which will in turn keep the transport operator in a competitive position. Thus, managers monitor the offered quality, as perceived from the point of view of their passengers, and point out the specific areas which require improvement. In practice, quality monitoring and benchmarking are two management methods, one addressing the objective, means and results, and the other comparing the improvement of an action undertaken, respectively. Both methods are acknowledged in the field of public transportation as tools to identify passenger priorities and needs, to measure passenger satisfaction, to assess service parameters and to indicate measures of improvement. Furthermore, passenger satisfaction is related to the perceived discrepancy between actual and ideal levels of service. Therefore, both perceptions and expectations of service are being considered, regardless of the management method (Nathanail 2008).

2.3.3.1 Travel time

The main inconvenience in intermodal travelling is caused by changing modes and waiting at interchanges, which represent barriers to 'seamless journeys'. Research literature suggests that these barriers, also known as the *interchange penalty*, dissuade people from making use of multimodal transport, particularly between different forms of public transport (Wardman and Hine 2000; Hine and Scott 2000; Crockett et al. 2004; Preston et al. 2006; Di Ciommo et al. 2009). The interchange penalty comes as a result of the extra time required to make a connecting journey while waiting for the connection to the next mode, which is also tied heavily to the risk of missing a connection and the perceived hindrance associated with the inconvenience and discomfort of the interchange. This penalty is expressed in higher weights that people assign to waiting and walking times in comparison with in-vehicle time. When people are asked to estimate the total travel time based on their perception of the individual travel time components, most often this leads to an overestimate (Rail Safety and Standards Board 2010) (Figure 2.5).

Figure 2.5 Waiting time at the interchange: St. Pancras Station, London, United Kingdom (a), and Oslo bus terminal, Vaterland, Norway (b). (Courtesy of Jan Spousta (a) and Julie Runde Krogstad (b).)

The interchange penalty can affect passenger numbers on public transport. In the Netherlands, the redevelopment of s-Hertogenbosch Station, following the opening of the new station in 1998, focused on three strategies to address the interchange penalty: accelerate (reduce the waiting time); condense (reduce access and egress times by locating facilities closer to interchanges); and enhance (make the interchange a more attractive place to users). This resulted in an increase in passenger numbers from 26,800 in 1996 to 40,100 in 2002 (Van Hagen and Peek 2003). Other measures include the reduction of access and egress times, in conjunction with that of the waiting time, as in the case of Moncloa in Madrid (HERMES 2011).

2.3.3.2 Punctuality and reliability

Punctuality and reliability in public transport are associated with service quality. Punctuality measures the deviation of the arrival time of a mode of transport from the scheduled one. Reliability relates more to the transportation networks and represents the ability of the operator to adjust the itineraries and schedules to accommodate unexpected incidents that create network disruption with minimal impact on service efficiency. Both elements affect user behaviour and, finally, the journey time.

For example, bus travel is considered less punctual and reliable than rail; therefore, travellers prefer the latter when arriving on time is important (Hine and Scott 2000). Even within the rail network, punctuality is perceived to be a barrier to making connections. It is worth noting that in the United Kingdom, the rail industry's punctuality targets are based on times of arrival at the final station. The result of this is that trains recorded as being 'on time' are possibly still late and miss connections at intermediate stations (Crockett et al. 2004).

2.3.3.3 Comfort and convenience

The necessity of seating and shelter while waiting is one of the issues to take into account. Shelters offering weather-protected seating or where all-round visibility is available increase people's feelings of security and, in turn, their comfort (Commission for Integrated Transport 2000). Plaza de Castilla in Madrid is one particular example of an interchange with this specific requirement in mind with a large marquee joining two separate passenger areas to prevent passengers' exposure to the elements (Consorcio Regional de Transportes de Madrid 2010).

Comfort is also linked to the stress or anxiety people feel at interchanges, such as the discomfort experienced in overcrowded areas. Overcrowding can cause stress, as it can make it difficult to move around an interchange, increase confusion about which way to go and consequently increase the risk of missing connections. Discomfort is also caused by 'forced' body contact at crowded places; however, this depends on the cultural

Figure 2.6 Overcrowding at King's Cross Station, London, United Kingdom. (Courtesy of Jan Spousta.)

background of the person. Crowding results in significant reductions in pedestrian convenience as movement speeds are restricted due to a loss of freedom to manoeuvre within the traffic stream. Since convenience is a significant factor, substantial overcrowding is likely to reduce people's propensity to use interchanges, or at the least, increase their perceived value of time spent there. In Madrid, the interchanges have undergone mobility simulation studies to ensure that movement around them is safe and comfortable for the user (Consorcio Regional de Transportes de Madrid 2010) (Figure 2.6).

2.3.3.4 Safety and security

Security is closely linked to comfort and is a concern, especially in darkness as people feel more comfortable in well-lit areas. Unaccompanied females and older users in particular perceive walkways and tunnels at interchanges as being less secure than platforms and booking halls when travelling at night. Passengers also have concerns about crime and overcrowding at busy interchanges, feeling they may be more at risk. However, they also feel at risk when interchanges are overly quiet or located in isolated areas (Figure 2.7).

2.3.3.5 Integrated servicing: ticketing and information

When interchanges involve more than one mode, considerations also have to be given to the implications of different timetables, ticketing systems and information provision (Wardman and Hine 2000).

The GUIDE (Terzis and Last 2000) project indicated that fares and ticketing policies have a great impact on public transport use. Multimodal

Figure 2.7 Long corridor for transferring at Paseo de Gracia interchange, Barcelona, Spain. (Courtesy of Lluis Alegre.)

tickets and other integrated ticketing systems (not separate tariffs for different modes) have a strong impact on passenger demand. Monzon et al. (2013) indicated that ticketing is a very significant factor for choosing travel patterns among different groups (by age, by trip purpose, but also by interchange). Establishing sufficient capacity for ticket offices or ticket machines is likely to be a cheaper measure compared with design changes for the overall interchange. Barriers connected to validating or buying tickets could be reduced by having integrated and/or electronic ticketing. Travellers will have less need for buying tickets at interchanges and this could lead to fewer queues and higher user satisfaction (Nathanail 2008).

Concerning information provision, the unpredictability of the journey experience is a significant problem (TfL 2009). Having information before travelling makes users more comfortable with their choice of journey, reduces stress and cost, and allows users to control their time spent at the interchange when changing transport modes. The information requirements of disabled passengers are more demanding than those of others as alternative route options may be more limited. Accurate and reliable real-time information tailored to the needs of disabled people is therefore vital, as is information prior to the journey. This can give users information not only about their journey, but also about the facilities available to them at an interchange and whether facilities such as lifts and escalators are out of service, thus enabling them to prepare and thereby reducing unpredictability and concerns about using interchanges. Although it is unlikely at the moment that a passenger could get integrated information about all timetables for all possible modal connections, some systems provide information about access on individual bus routes, for example, whether buses have low, easily accessible floors for the disabled or people with small children, and luggage or wheelchair access. It is also possible to obtain information

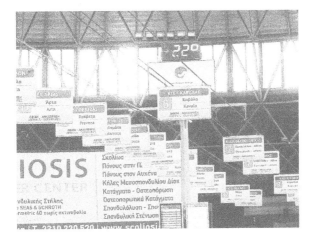

Figure 2.8 Nonmultilingual information provision at Macedonia Coach Terminal, Thessaloniki, Greece. (Courtesy of Jan Spousta.)

about whether bus routes have Wi-Fi availability on board and accept smart card tickets, giving users information in advance about how they can use their time and how they can purchase tickets. Audio-visual assistance systems such as those indicating the arrivals of buses and trains are available in different technologies in cities such as Madrid, Prague, Dresden and Linz, but are still not very widespread throughout Europe (Cré et al. 2008). International travellers would benefit from multilingual information and thus will make easier use of multimodal transport (Figure 2.8).

A range of different information services have been identified and reviewed for Rail Research UK (Preston et al. 2006; Crockett et al. 2004). They conclude that these developments are promising and cite research showing that such services have the potential to encourage modal shift. However, this is another aspect of multimodal travel that is not currently fully taken into account in the transport models used for passenger demand forecasting, making it more difficult to develop a business case for investment in improved information. Further research is needed to quantify the impacts of information systems on passenger demand so that forecasting methods can take this into account. It will be necessary to ensure that information and journey planning services are designed and targeted to meet the needs of different users and journey purposes, in particular those who are not currently public transport users.

2.3.3.6 Facilities

There is some research evidence to support the idea that people have a 'travel time budget', which not only limits the total amount of time people are prepared to spend travelling, but also implies that people prefer to have a minimum amount of travel. Metz (2008) argues that, across the world, people's average travel time per day has remained fairly consistently at about an hour. Reasons for this include the need to allow time for other activities, such as childcare, shopping, leisure and so on, accommodating at the same time the need to travel. An interchange can make the modal shift easier and faster (avoiding missing connections); it can also enable other activities while waiting for the modal change. Therefore, waiting time would not count towards the journey time and the individual's travel time budget, a fact which would increase passengers' willingness to undertake that journey.

Providing people with access to points of interest such as shops and places to eat while they wait at an interchange lowers their value of time, makes them feel more comfortable and reduces boredom. Facilities close to boarding areas are preferred. In the Netherlands, the Dutch Railways are focusing on enhancing the users' time at the interchange. This is achieved through improving their experience with amenities such as shops and making the interchange a more enjoyable place to be with infotainment available and visual stimuli such as lighting, art and greenery (Van Hagen and Martijnse 2010).

In Madrid, following the construction of an intermodal exchange station (IES) at Avenida de América, more people used multimodal transport because they no longer had to wait for buses in the street, but could look around shops instead. The IES also improved the look of the surrounding area, encouraging more people to use it (Vassallo et al. 2012).

2.3.3.7 Accessibility

Interchanges involve physical effort to move from one platform to the other, and may require walking long distances and possibly the use of stairs, which are especially inconvenient for people with special mobility needs, such as the elderly and those travelling with children or luggage. At interchanges, the elderly are less willing to accept changing platforms because it is inconvenient due to their reduced mobility, and it is often not possible for them to determine, prior to the journey, whether lift facilities are available. People with luggage also have a similar concern as their luggage can act to reduce their mobility with the risk that they will not be able to make a connection. This concern is higher for short connection windows. Interchanges can involve several places where luggage handling is necessary, which can dissuade people from using multimodal transport (Wardman and Hine 2000) (Figure 2.9).

Improved accessibility reduces not only the interchange time period but also the travel time of the whole journey and improves the usefulness of the time as well. Improvements in access can be a very cost-effective way to improve perceived journey time (Passenger Focus 2011).

Although satisfaction with accessibility has increased slightly over the last decade, for example, in the Netherlands and in Germany (Brons et al.

Figure 2.9 Access for all: Paseo de Gracia interchange, Barcelona, Spain. (Courtesy of Lluis Alegre.)

2009; Woldeamanuel and Cyganski 2011), it has not increased as much as overall satisfaction with public transport. The importance of accessibility has increased over time. The combination of these developments has resulted in a steady increase between 2001 and 2005 of the negative impact of accessibility on overall satisfaction with rail travel. While it is encouraging that the satisfaction with access is improving over time (the mean satisfaction score for each year is statistically different from that of other years), the higher importance passengers assign to this dimension, the more continuous improvements are required (Brons et al. 2009).

Tsami et al. (2013a) investigated the accessibility level in Thessaloniki Railway Station's urban interchange under seven different accessibility scenarios, and results showed a willingness to increase multimodal travelling if interventions could improve the connections of the interchange with the rest of the public transport network. This would also improve overall travel time.

Physical barriers for people with disabilities are primarily related to boarding and alighting public transport modes at interchanges; however, such barriers and crowding may restrict movement at the terminal, the platforms and on board the vehicles. As the main mode of transport for many disabled people is walking (or a wheelchair), improvements are required to improve access outside the interchange through measures including clearing obstacles and introducing tactile paving. This will not only enhance the interchange experience but also improve the whole journey. A number of interchanges in Europe are notable for their barrier-free access and tactile design for users with reduced mobility. These include Berlin Railway Station, Gare do Oriente in Lisbon, Linz Central Station, and Moncloa in Madrid (KITE project 2008; HERMES 2011).

2.4 INTERCHANGES: THE SEAMLESS SOLUTION

Throughout recent decades, social and economic opportunities have contributed to a growth in journeys, causing a significant increase in the congestion of urban areas and environmental deterioration. While satisfying the demand for mobility is a key determinant factor for citizens' quality of life, transport has started to pose a threat for modern societies.

Urban public transport can provide a feasible alternative to individual car transport; however, the availability of opportunities for direct journeys when using public transport is limited and the majority of trips require a combination of modes, thus also including a shifting from one mode to the other. An incentive to increase public transport utilisation is to integrate such operations at interchange facilities (Banister 1999).

2.4.1 Interchange principles

The role of an interchange is to enable passengers to change from one route or mode to another, taking into consideration that they may need to exit the system or wait for their connection. Although larger interchanges are being designed to offer a variety of commercial and retail opportunities for travellers, it is important to note that the core design of the interchange should be focused on transport transfers. The City-HUB project identified guiding principles that can be applied to any interchange and set the basis for those who plan, design or upgrade new or existing facilities (Table 2.2).

2.4.2 Interchange goals

When assessing the performance of intermodal passenger transportation, a set of indicators may be used to address the different aspects that form the physical, service and institutional interfaces.

In terms of the physical interface, the supply side performance and the terminal properties are assessed. *Supply side performance* is connected to energy use, investments, performance and efficiency in the utilisation of resources, financial performance, social standards and transport volumes/flows. These issues may also be relevant throughout the journey. *Terminal properties* are aspects of the specific terminal or transport leg interface, capturing the design, location, accessibility, scope of services offered, signage, space and capacity offered, as well as the technology and equipment possessed.

Institutional interfaces are addressed through the assessment of the *organisational and institutional structure*. This structure refers to the role of and relations between organisations (stakeholders), for example, ownership, responsibility for infrastructure and operation, and the institutions that affect these organisations, such as regulatory and financial structures. These issues apply throughout the journey, meaning at both the access and

Table 2.2 Guiding principles

Principle	Content
Clarity of purpose and functions	Clear layout for the interchange and its functions.
Accessibility	Physical accessibility, including step-free and barrier-free access to all, and a good location of the facility and so on are two aspects to consider.
Legibility	Legibility refers to a legible design environment, which make navigation easy and movement within the interchange seamless. Key elements of legibility include layout, lighting, surfaces and materials, and finishes and furniture.
Governance	The more complex the interchange becomes, the greater the need for the management and ownership structures to be formalised and include users' participation in planning and operating an interchange.
Financing and business models	Function, size and context of the facility determine the choice of business model to be adopted for an interchange. The majority of these models separate the capital cost from the operational costs and revenues. This is feasible when the interchange is planned, since initial capital investments comprise the largest proportion of the costs related to the life cycle of the interchange.
Regulations and legal aspects	National and local regulations influence aspects of the design and development of an interchange, and determine which business models are applicable. Basic levels of safety and security must be provided, as well as modal-specific requirements.
Adaptation	An interchange should be a dynamic component of the urban transport network, adapting to the changing environment, composed of user growth and business, social and technological development.

egress points, along with travelling and interfaces/terminals. The organisational and institutional structure is affected by *policy objectives and measures* that affect the entire transport system, including objectives connected to modal split, environmental effects, efficiency and safety, as well as measures that initially can be divided into broad categories such as economic/financial, legal and physical/infrastructure.

Information and fare interfaces are related to the *level of service*, which represents the quality and cost that are delivered to customers, including concepts such as relations with customers, comfort, cost, flexibility, frequency of services, information delivered, reliability of service, safety and security issues, integration of services and integration of fares/tickets, as well as time use and efficiency in the operations. Level of service may be considered at different assessment levels and on different legs within a journey.

Table 2.3 depicts the respective enablers that apply in each of the aforementioned dimensions and their respective goals.

Table 2.3 Key enablers and recommended 'goals' for promoting future intermodal mobility

Intermodal mobility enablers		Policy goals	Organisational and institutional structure goals	Supply-side performance goals	Terminal properties goals	Level of service enhancement goals
Physical	Enhanced 'green transport' usage.	Reduced greenhouse gas emissions for all trips attracted by the interchange.	Fair and equal access for all.	Reduction of motorised traffic.	Facilitation of 'green modes' use in terms of accessibility.	Security and safety of 'green mode' travellers.
	Interchange access and egress.	Incentives for public transport integration.	Separation of the infrastructure-owning entities from transport operators.	Maximisation of 'green energy' usage.	Ease of accessibility to the interchange.	Maximise the seamlessness of complete door-to-door travel.
	Interchange mobility.	Standards for harmonised interchange design for users with reduced mobility.	Promote sharing systems and temporary intertrip cooperation of service providers.	Use of advanced guidance and safety systems.	Infrastructure integration.	Minimise total interchange time and delays.
Service and information	Scheduling coordination, payment integration – simplification.	Harmonisation – integration of fare and fare collection systems – use of IT to simplify payment and issuance. Door-to-door integrated travelling planning.	Revenue sharing system. Consideration of technological, infrastructure capacity, fiscal and legal dimensions.	'One-stop' shop servicing.	Optimise interconnection between platforms, signage and instructions for the integrated system, ticket issuance.	Incitation for integrated ticket issuance through price reduction. Minimisation of transfer time. Optimising use of interchange time.

	Comprehensive travel and trip information.	Integrated operation of all user information systems by all involved networks.	Harmonisation of information and information visibility.	Reduce complexity.	Clarity and simplification of information points.	Highly accurate and reliable information provision.
	Personal navigation support.	Inclusion of environmental and other objectives in intermodal route planning.	Provide a common platform for information retrieval and sharing.	'Equilibrium' in network utilisation.	Provision of all necessary information and communications technology (ICT) infrastructures.	Provide clear and efficient decision-support to individual users satisfaction.
Institutional	Institutionalised collaboration of stakeholders.	Public transport connection prioritisation. Deregulation of services.	Explicit definition of roles, jurisdictions and obligations. Harmonised regulatory framework for interconnections.	Balance between flows and infrastructure resources used.	Integration of terminals with the rest of the network. Reduction of carbon fuelled cars. Guidelines for infrastructure and accessibility.	Compensation for late or early arrivals by adjustment of sequential trip legs.

Chapter 3

Interchange place

*Floridea Di Ciommo, Andres Monzon and
Ana Barberan*

CONTENTS

Once the need for interchanges is clearly stated, it is necessary to mark a clear roadmap to achieve people-focused, seamless and efficient transport interchanges. This is a multifaceted task that should consider multiple aspects within the complexity of urban areas. Developing or designing interchanges is an intensive and complex process that involves different stakeholders, produces various impacts and also creates new city landmarks (Figure 3.1).

Madrid Regional Transport Authority redefined a transport interchange station as not only an 'area whose permanent purpose is to facilitate the interchange of people between various modes of transportation' but also an 'area whose purpose is to minimise the inevitable sensation of having to change from one mode of transportation to another' (Consorcio Regional de Transportes de Madrid 2010). Di Ciommo et al. (2014) show that users identify the improvement of city-hubs with the quality of time spent inside them. The current challenge of interchanges is to facilitate transfer from the use of private motorised vehicles to the shared use of cars (i.e. car sharing or carpooling), to the use of public transport and non-motorised modes. It

Figure 3.1 Interchange functions: Stratford Interchange, London, United Kingdom. (Courtesy of Jan Spousta.)

is, in a certain way, a planning principle. The pivot of intermodal transport when designing interchange spaces will be a comfortable and practicable connection by constructing platforms, integrating information systems, installing bike and ride options and defining pedestrian flows around an interchange. Travel intermodality is not only a planning principle, but also a policy that aims to provide passengers with a seamless journey using different modes of transport in a combined trip chain.

For decades, transport investments were based on the contraposition between public and private modes of transport. Today, when increasing urban sprawl and related low residential density have already increased car dependency, even in Europe (Lucas and Jones 2009), mechanised trip dependency is a fact. Because of current land use patterns and urban sprawl, it is quite impossible to travel less, as wished (Banister 1999); therefore, the only option for policy makers is to plan for better travel. More than 20% of current commuting trips in Europe could be intermodal, and between 10% and 20% of travel time would be spent in intermodal transfer. In this context, interchange infrastructures are part of innovative transport planning measures oriented to achieve more sustainable urban mobility patterns.

In this new kind of intermodality, a key role will be played by providing information about transport-related attributes such as travel time, travel costs and environmental quality standards. All these elements might be seen not only as a service to users but also as an instrument to change their travel behaviour. While economic theory suggests that individuals base choices on the attributes of the choice set (information content), the way that information and services are presented (information and space contexts) also has a strong effect on travellers' behaviour. *Choice architecture theory* shows that small features incorporated into the environment of choice making act as 'nudges' to help individuals overcome cognitive biases and to highlight better choices for them – without restricting their freedom of choice. This literature suggests examples of how influencing travel behaviour through the design of transport infrastructure helps to promote desirable travel options (Avineri 2012).

Bertolini (2006) discussed the previous consideration of a multimodal passenger interchange as a 'non-place'. The objective of the City-HUBs book is to define guidelines for transforming a 'non-place' into a pleasant meeting and transport 'interchange place'. The main benefits of urban interchanges relate not only to time saving, but also to a better use of waiting time. They are oriented to increase potential combinations of both private transport (i.e. bike, car and ride sharing) and public transport modes (i.e. bus, rail and metro), urban integration and land use and improved operational business models.

On the one hand, a transport interchange includes features related to its internal workings (way-finding, opportunity space, distances between modes, etc.), and on the other hand, the interchange is immersed in a relationship with its urban environment where each is affected by the other. Using some key aspects of the interchange, the aim of this chapter is to provide a typology for classifying interchanges.

3.I IDENTIFYING A TYPOLOGY OF INTERCHANGES

According to Grémy and Le Moan (1977), 'to develop a typology is to distinguish, within a set of units (individuals, groups of individuals, social events, social environments, etc.), that can be considered homogeneous from a certain point of view'. The content of this notion of homogeneity is usually based on a similarly defined subset of features to describe the studied units. A typology must meet two additional requirements: completeness and exclusiveness of type. In the case of transport interchanges, the use of a typology is motivated by the impossibility of reaching a single model of understanding how an interchange works.

In particular, in the case of the interchange, the procedure we proposed for constructing a typology was divided into two phases. The first phase includes the analysis of basic aspects of *functions and logistics* of an interchange which give an idea of the order of magnitude of the *interchange size*. The second phase makes use of this size together with the surrounding *local constraints* of the interchange to produce the *interchange typology*.

3.2 THE INTERCHANGE PLACE IN THE CITY

The empirical work on interchange typology was based on a qualitative survey undertaken at 16 selected interchanges through interviews with practitioners, transport planners from transport authorities, transport operators or those in charge of interchange business development. This information has been complemented by the detailed analysis of five pilot case studies. The analysis focused on the functions and logistics aspects, including daily passenger traffic, types and number of transport modes, services and facilities and the location in the city, as well as local impacts.

The analysis of the collected information and opinions identified two *dimensions* or groups of aspects that interact to define the needs of the *interchange place* and consequently the size of the building and its characteristics (see Figure 3.4).

The first group of aspects (*Dimension A*) is related to the internal *functions and logistics* of an interchange, including transport elements of the interchange and the services and facilities necessary to fulfil the transfer functions properly. This dimension determines the size of the terminal building.

The second group (*Dimension B*) includes the external aspects of the city environment that affect how the building could be in reality. This dimension includes the location of the interchange within the city and whether or not the interchange plan is in conflict with the existing land uses in the surrounding area.

3.2.1 Dimension A: Functions and logistics

The first group is related to the functions and logistics aspects, including demand, modes of transport and services and facilities. They are not independent and can be defined as follows:

- *Demand*: The number of passengers is the first aspect to define the interchange size, as this aspect determines the need for space and access characteristics. Three levels of this aspect are described: (1) less than 30,000, (2) between 30,000 and 120,000 and (3) over 120,000 passengers/day.
- *Modes of transport*: The second aspect is related to the modes of transport included in the interchange and their degree of importance. Three different levels resulted from the qualitative analysis: (1) interchanges with buses as the dominant mode of transport, (2) interchanges with rail as the dominant mode of transport and (3) two or more public transport modes or different lines of the same mode jointly (Figure 3.2).
- *Services and facilities*: This aspect is related to the number and quality of services and facilities located at an interchange. Services and facilities will depend on the volume of passengers transferring in the

(a)   (b)

Figure 3.2 Bus- and rail-based interchanges: Méndez Álvaro Coach Station (a) and Ilford Railway Station (b). (Courtesy of Samir Awad (a) and Jan Spousta (b).)

interchange. It could have three different levels, including: (1) a few kiosks or vending machines; (2) a few retail shops, cafés or food facilities for travellers; or (3) a shopping mall integrated with the interchange.

3.2.2 Dimension B: Local constraints

This dimension has three interrelated aspects to consider in an aggregated way. The first is related to the relative location of the interchange with respect to the main local demand attractions. However, the building of the interchange is also affected by the kind of activities developed around it. If the city considers the transport interchange as part of its urban development plan of the area it is even more important. The description of these aspects is as follows:

- *Location in the city*: The geographical aspect of an interchange is related to its location in the city. The qualitative analysis of 21 interchanges shows that urban interchanges could be classified as being located in: (1) suburban areas; (2) at the entrance to the city, where major public and private transport modes connect the outer and inner city or a different part of the city; or (3) in the city centre, where people interchange mainly for moving inside the city or within the peripheral urban areas.
- *Surrounding area features*: The activities located in the surrounding area could support or become a limitation to the activities associated with the interchange. Green areas or heavy industry could be a limitation, but a large commercial centre or sport field could foster the use of the interchange for access to transport and the use of services inside (Figure 3.3).
- *Integrated development plan*: The interchange infrastructure could be part of a local development plan to encourage economic and urban

Figure 3.3 Surrounding area features, a key aspect. (Illustration by Divij Jhamb.)

development, especially when urban regeneration policies are needed. Commercial development, new housing and offices are more likely to occur when an interchange is integrated into a development plan. The consequent involvement of local government will be required when the interchange infrastructure is integrated into a local development plan.

3.2.3 Interchange place size and typology

The left-hand side of Figure 3.4 shows the causal relationships between the two dimensions of the interchange place: functions and logistics, and local constraints. The functions and logistics aspects define the physical size of the interchange (i.e. the building structure of the terminal) and its form (i.e. design). Therefore, interchanges are characterised by a flow of travellers, the number of public transport modes that serve the interchange and their associated retail and commercial outlets among other services and facilities. All together, these attributes will determine the need for space and the setting for all these activities in an ordered and coordinated way. The interchange size would be classified as *small*, *medium* or *landmark*.

A *small* interchange place is characterised by low passenger flows, a small number of transport modes that service the interchange and only a few kiosks or basic facilities inside. A *medium* interchange place is characterised by an intermediate flow of passengers, a considerable number of transport modes and some retail and food facilities for travellers. A city *landmark* fits with a higher flow of travellers, a complete range of public and private transport modes and significant retail and/or an integrated shopping mall.

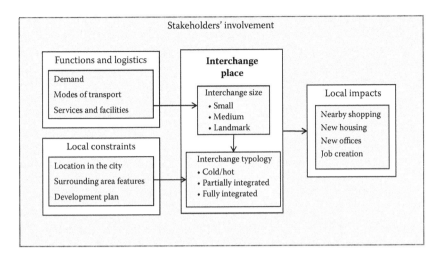

Figure 3.4 Urban transport interchange place.

However, the amount of space dedicated to the interchange is also affected by the local constraints, which determine the particular features of the building design. In the city centre, the interchange building will be more constrained than in a suburb where the availability of space allows the interchange infrastructure a wider area. The combination of the two dimensions of an interchange place could define the typology of an interchange as: *cold/hot*, *partially integrated* and *fully integrated*.

3.2.4 Local impacts of the interchange

Let us consider the right-hand side of Figure 3.4. The complete interchange with several activities located inside also has *local impacts* that include the consideration of the economic and land use effects in the vicinity of the interchange. There is a clear interrelation between the interchange's size, local impacts and typology. This relationship creates a dynamic causal chain.

The considered variables for these local impacts are nearby shopping, new housing, new offices and job creation. They could be detailed as follows:

- *Nearby shopping*: Passengers using the interchange provide a business opportunity for the area where the interchange is located. This has been identified as having a clear effect on the activities of shops in the surrounding areas and the creation of new opportunities to serve travellers' needs.
- *New housing*: Interchanges could have an impact on the local economy and land use. New housing can be constructed on top of or near to the interchanges. New housing development could be possible

when land use constraints are relatively low and vacant land is available for use nearby. When the interchange is part of an urban regeneration policy, this land could be designated as greenfield or brownfield.

- *New offices*: New offices can be placed on top of or near the interchange. Office development could be possible when land use constraints are reduced and an area of greenfield or brownfield land is available locally.
- *Job creation*: A key factor for evaluating the local economic impact of the interchange is related to job creation. No statistical study is available for estimating the number of new jobs at an interchange scale. However, our qualitative analysis shows that in some cases of interchange development, job creation is observed. This element requires the involvement of local government and the owners and businesses in interchanges.

3.2.5 Stakeholder involvement

When interchange construction or refurbishment produces new housing or offices, local government, owners and business stakeholders will be involved in defining policy goals and mapping out the opportunities for the interchange management.

When the interchange status analysis detects the presence of all or some of these local impacts, stakeholder involvement is needed and a more complex business model for managing and operating the interchange will be required.

Stakeholder involvement is based on the following:

- *Local government*: Local development plan and coordination of land uses in the surrounding area.
- *Developers and businesses*: When some local impacts exist on economic activities, such as nearby shopping, new housing, new offices and job creation.
- *Users* are always included, as well as *operators*, as they are closely associated with the functions and logistics aspects.

3.3 METHOD FOR AN INTERCHANGE TYPOLOGY

The key aspects of an interchange characterise it. By comparing them, we can propose a typology of interchanges. The interchange place is determined by the functions and logistics aspects (i.e. travel demand, modes of transport and services and facilities) and the local constraints (land uses and their territorial distribution and location in the city). Both interchange dimensions can be analysed separately, as shown in Tables 3.1 and 3.3.

Table 3.1 Functions and logistics aspects influencing interchange size

Dimension A aspects	Levels	Need for space in the interchange	Score level
Demand (users/day)	<30,000	Low	1
	30–120,000	Medium	2
	>120,000	High	3
Modes of transport	Dominant – bus	Low	1
	Dominant – rail	Medium	2
	Several modes and lines	High	3
Services and facilities	Kiosks, vending machines	Low	1
	Several shops and basic facilities	Medium	2
	Integrated shopping mall with all facilities	High	3

The three aspects of Dimension A are presented in Table 3.1. The requirements for space in the interchange building are categorised in three levels: low, medium and high. We have assigned a score of one to three to each. These scores represent the relative space required by the different levels. Scores could be assigned in this simple way or be determined through ad hoc surveys among stakeholders. Such surveys could provide scores for each level and also the relative weight for each of the interchange aspects. In other words, through surveys, the stakeholders could state if, for example, the demand level should have a higher importance than the number of services and facilities. First, it is necessary to decide what level of space is required for each aspect of the interchange under study, as presented in Table 3.1. The appropriate size of an interchange can be settled by adding the scores for each of the three aspects of Dimension A: demand, modes of transport and services and facilities. The results are shown in Table 3.2, which proposes that the size of the interchange could be small, medium or landmark, according to the score ranges.

Once the size of the interchange is clearly fixed, we have to consider how the local constraints, included in Dimension B, could affect the interchange development. The aspects of Dimension B influence the needs for

Table 3.2 Interchange place size

Total score of Dimension A aspects	Interchange place size	Description
3–4	Small	Low level for all functions and logistics aspects or at most one medium level.
5–7	Medium	Combination of levels for the three aspects that provide an intermediate average.
8–9	Landmark	High level for at least two of the three aspects that require large-scale building.

Table 3.3 Local constraints aspects influencing interchange typology

Dimension B aspects	Levels	Upgrading level	Value
Location in the city	Suburbs	Less	−
	City access	Neutral	o
	City centre	More	+
Surrounding area features	Non-supporting activities	Less	−
	Supporting activities	Neutral	o
	Strongly supporting activities	More	+
Development plan	None	Less	−
	Existing	Neutral	o
	Existing and including intermodality in the area	More	+

space associated with Dimension A aspects. Therefore, the need for space determined according to Table 3.1 scores could be more, or less, due to the influence of local constraints. The same improvement customising the values and weights, as before, could be undertaken if specific surveys to stakeholders were carried out.

Table 3.3 shows how the different aspects of Dimension B would modify the size and characteristics of the interchange building. It proposes a value to modify the characteristics of the building: that is, whether it should be upgraded or not. The negative value means lower impacts and less constraints and the positive one signifies the need for more integration in the surrounding area. As an example, passenger terminals located in a suburban area, or at the entrance to a city, could be easier to design and to integrate than interchanges with the same flow of travellers and transport modes but located in the city centre.

The values given in Tables 3.1 (Table 3.2) and 3.3 define the interchange place typology. This typology includes the physical elements (size and type of building) and the qualitative aspects included in the local constraints considerations. The combination of both dimensions gives rise to the most suitable interchange typology. We propose the following three types of interchanges:

3.3.1 Hot or cold interchange

Depending on the climate, this category refers to the case where the interchange operates in an open-air environment. Generally, it is located in a suburban area and the type of activities around are not very significant. These interchanges can only offer a limited number of services, including ticket vending machines, kiosks or snack vending machines, with a limited degree of intermodal integration. At these interchanges, there can be a low to medium flow of passenger numbers, but the services provided will be

Figure 3.5 Cold/hot interchange. (Illustration by Divij Jhamb.)

limited by the facility itself. The analogy referred to is similar to fuelling stations which are unmanned, offering basic services but with limited supporting infrastructure. Such interchanges are not part of the urban development plans of the area but rather a way to solve the problem of transfer in an ordered way (Figure 3.5).

3.3.2 Partially integrated

The partially integrated type involves a wider selection of services that are provided within the interchange but not necessarily providing all the services required by passengers. Normally, it is not located at the centre of the city, but it does have compatible types of land uses around it. The usual features are shelters and covered platform areas, shops and other small businesses. Typically, interchanges built in connection with existing terminals, such as train or bus stations, fall into this type of operational model. There, the existing facilities set limitations on the number of additional services, although in some cases these models have worked well in terms of revenue generation, when former station spaces have been converted into additional service areas. The selection of services available should not only be restricted by the availability of space but also be demand-driven. Limitations to such operational types come from the existing physical infrastructure; for instance, a railway station, even when refurbished, can only allow a limited space for retail, restaurants and other facilities (Figure 3.6).

Figure 3.6 Partially integrated interchange. (Illustration by Divij Jhamb.)

3.3.3 Fully integrated

Fully integrated models are most common in newly developed interchanges, as the setting requires the design and construction of facilities so that the integration and mobility of passengers can be designed optimally. Located in the city centre, or at the access to the city, they are part of an integral development plan. In integrated interchanges, the driving force should be seeking mutual benefits from the integration of transport activities with commercial ones (Figure 3.7).

3.4 FURTHER USE OF INTERCHANGE TYPOLOGY

Interchange analysis is a complex activity that includes transport networks, services and urban aspects. Nevertheless, the analysis of each key aspect of the 21 existing interchanges studied allowed a typology to be developed related to the interchange logistic functions and size aspects, and local constraints, which will be useful in determining the interchange place design, the stakeholders' involvement and the local impacts. The interchange typology demonstrates how to check the interchange status and identify the type of interchange being dealt with.

Chapter 4 will present the City-HUB life cycle focusing particularly on the governance process and stakeholders' identification and involvement.

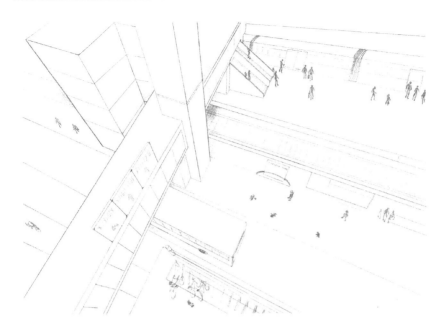

Figure 3.7 Fully integrated interchange. (Illustration by Divij Jhamb.)

Each type of interchange deals with the organisation of services and facilities at an interchange. The facilities provided at any interchange could correspond to a minimum level of services or scaling up of additional facilities to meet users' requirements of a larger interchange. This aspect will be analysed in detail in Chapter 5. The type of interchange depends on the level of traveller demand. When the services and facilities inside an interchange are higher, or lower, than the typology suggests, users' satisfaction will be higher or lower as well, as shown in Chapter 6. Both the building size and design, the expected local impacts and the degree of stakeholders' involvement will determine the business model to adopt for managing each type of interchange. These will be analysed in Chapter 7.

Chapter 4

Insights

Interchange management and governance

Ricardo Poppeliers, Floridea Di Ciommo,
Odile Heddebaut, Marko Nokkala and Tuuli Järvi

CONTENTS

There are various governance models for managing public facilities. Di Ciommo (2004) and Martens (2007) define three ideal models of governance that could help us to explain the role of various interchange actors in different contexts. The first one is the *coordinative model* that includes continuous efforts in coordination between various parts of the governing body. In this model, the governing body and the governed create the framework, but only those actors who have the authority to decide are part of the governing body. They are able to articulate the public interest, to determine the need for strategic planning and to select the best policies and programmes. The second option is *governance through competition*, which takes its origin from political theory, market economy and the pluralist model of democracy. In this model, governance is seen as a competition between actors of different interests. The Spanish case of Moncloa seems to be explained by this governance model. The third could be called *communicative planning*, where the main idea is that governance should be a process of agreement between all stakeholders involved. This model seems to be appropriate for the good management and operation of a complex transport infrastructure such as an interchange.

Interaction between relevant actors involved in each of the steps of the interchange life cycle results in established governance structures. Figure 4.1 shows the governance process alongside the *City-HUB life cycle*. There are

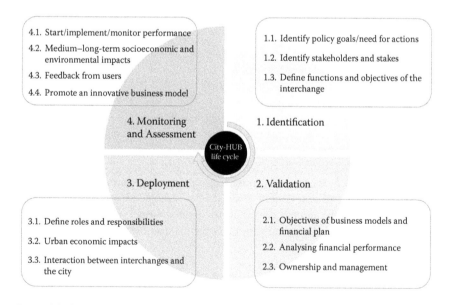

Figure 4.1 Governance process alongside the City-HUB life cycle.

basically four steps which are shown in Figure 4.1 and will be further elaborated throughout this chapter.

4.1 IDENTIFICATION

4.1.1 Identify policy goals and needs for actions

At a time when transport is considered key for urban sustainability and social inclusion in the political agenda, and decision makers have started implementing a wide range of policy measures towards sustainable urban transport, we are confronted with an increasing need to define measures to influence travellers' behaviour, to increase the public transport share and to develop and deploy innovative, sustainable mobility modes. The quality, accessibility and reliability of transport services is gaining increasing importance, inter alia due to the growth of the population, urban sprawl and the need to promote public transport through intermodal trips. Faced with the increasing need for mechanised trips, the only realistic alternative to the car is the intermodal trip. The intention of a good interchange will generally be to improve the quality of public transport services and support seamless door-to-door travel. Nowadays, an interchange is more than just a simple node in a network; it has many elements.

In this context, urban transport interchanges play a key role as components of public transport networks in cities to facilitate the links between different transportation modes, particularly the connection of bus services to the subway, metropolitan railway system, biking and kiss-and-ride systems.

The benefits of urban interchanges mainly relate to time saving, better use of waiting time, urban integration and improving operational business models, and have a threefold goal:

- Firstly, to facilitate the users' transfer between two or more public transport modes by reducing the transfer time and increasing comfort. Users' travel time is reduced since they do not have to change location and the signs available inside the interchange facilitate easy modal connections. Users' comfort is increased as they can use the shops and restaurants available at the urban interchange making the transfer time more pleasant.
- Secondly, to coordinate public transport services through the information services provided at the interchange facilities. Seamless services are improved by the coordination of timetables and the necessary information and orientation at the interchange premises.
- Thirdly, to use urban space more efficiently. For example, before the existence of an urban interchange, space occupied by urban regional buses that used to stop on the street for passengers to disembark

reduces the capacity of the urban road network. This can be reorganised to be more efficient by making use of the interchanges. Another example is to design an interchange in such a way that several urban areas have direct access to both sides of the interchange. The organisation of a well-designed urban interchange can improve the image of the urban area substantially, may increase the value of adjoining properties and foster more and better local business.

Besides accessibility improvements, management and innovation, an efficient use of interchanges should also be depicted.

Therefore, policy goals are expressed in terms of welfare for citizens, influenced by effective transport policies fulfilling the needs of the citizens, but also by optimal strategic spatial planning leading to an even higher impact on welfare. An interchange station represents a way to contribute to the internalisation of negative environmental externalities and to increase social inclusion. The causal loop determining the need for an interchange includes the following steps:

- The increasing sprawl of the city produces an increasing dependency on mechanised trips.
- These mechanised trips could be realised by car or by intermodal trips.
- Intermodal trips are characterised by penalties and inefficiencies, which could be reduced through the construction of an interchange.

The European White Paper (European Comission 2011) emphasised that a competitive transport system should be realised with an important role for collective transport. The integration of different passenger modes is important to provide seamless multimodal door-to-door transport.

4.1.2 Identify stakeholders and stakes

In this step, the manager of the policy cycle should identify his or her own stake as well as (the stakes of) the relevant stakeholders, which are necessary to engage in the governance process in order to reach the identified policy goals. This is the start of the governance process.

4.1.2.1 Why involve stakeholders?

The endorsement of proposals to construct, expand or change an urban multimodal transport interchange by the stakeholders concerned will give decision makers and funding bodies confidence that the proposals can be implemented successfully. The real issue, however, is not about whether or not to involve these groups, but with what aims and by what means stakeholder involvement should take place. As a minimum, there is a need at a technical level to know about how travellers, other transport decision makers and other stakeholders will respond to the various measures that might be included in a transport interchange. There are a number of reasons for carrying out more than the minimum requirements for public consultation (Chartered Institution of Highways & Transportation 1995):

- An interchange could have a major impact on areas where people live, work and carry out their daily activities, so it is right that the general public is involved in the process.
- Local people can usually help in identifying controversial issues and bring in their local knowledge about existing problems as well as ideas for resolving them, which can be helpful for formulating plans.
- Plans are unlikely to be adopted by the local authority or accepted by the government if there is evidence of strong public hostility.
- Plans will only succeed when they are implemented if businesses and the general public are willing to accept the proposals.
- Establishing a process of dialogue between the public and professionals helps to make both parties become more aware of the issues and options available, which will assist the later stages of implementation.

Good stakeholder engagement reduces conflicts, results in better planning outcomes and most importantly allows communities to have an influence over the future shape of the places where they live.

4.1.2.2 When to involve stakeholders?

There are five possible stages in the interchange planning process in which stakeholders may be involved:

1. The setting of objectives/goals for the interchange at the outset of the process.
2. Identifying current and potential future problems (need for actions).
3. Developing ideas for measures/facilities to be associated with the interchange.
4. Indicating levels of support for different proposals.
5. Deciding on the preferred plan for the interchange.

It is important at the outset to clarify the aims and limits of stakeholder engagement, so that suitable techniques can be identified and confusion about roles can be avoided. A 'Stakeholder Involvement Plan', prepared in advance of the process, can help.

4.1.2.3 Which stakeholders to involve?

A stakeholder is defined as any individual, group or organisation affected by, or able to influence, the proposed project and its implementation.

Table 4.1 Typical stakeholder groups

Transport actors	Government/ authorities	Local communities/ neighbourhood actors	Business and commercial	Other (academics, stakeholders specific to the context)
Modal operators (public and private) Other related transport service operators, e.g. taxis Car/bike sharing groups (if their services were planned into the interchange) Other mobility providers	Local government Politicians Traffic/transport police/ emergency services Health and safety executives/local hospital representatives Neighbouring town council representatives	Faith leaders Local community organisations/ groups (e.g. sports groups, scouting movement, etc.) Transport user groups Representatives of marginal/minority or hard to reach groups (by culture/disability/ age/gender, etc.) Local environmental groups	Local chambers of commerce/ business associations Retailers or retail/ commercial groups that will use or rent space in the interchange for commercial purposes Local major employers	Universities and educational/ training establishments Special interest groups (e.g. environmental groups) Experts and consultants Financial actors

This includes the general public as well as businesses, public authorities, experts and special interest groups. The extent to which stakeholders are affected by a project and able to influence the process may differ. Wefering et al. (2013) make a distinction between:

- Primary stakeholders – those who are (positively or negatively) affected by the issue.
- Key actors – those who have power or expertise.
- Intermediaries – those who have an influence on the implementation of decisions, or have a stake in the issue (such as transport operators, NGOs, the media, etc.).

It is crucial to take account of these differences in selecting the right stakeholders as well as in choosing the best mode for stakeholder engagement and stakeholder management. In addition, the various functional areas should be considered in the selection process. Table 4.1 shows some typical stakeholder groups clustered into five categories.

It is not the aim here to fill in the broad picture covering all types of stakeholders. The intention is to analyse the most important groups of stakeholders capturing primary stakeholders and key actors, as well as intermediaries and their perspectives. Based on the case studies, the following stakeholder groups have been selected:

- *Interchange users:* Passengers' requirements depend on their mobility needs, but generally the benefit can be expressed in time, cost savings and convenient travelling (economists often express this as generalised costs). This relates not only to efficient changes between modes, but also to comfort. Not only do mobility services facilitate seamless travel (such as bicycle stands, parking lots, bus lanes, etc.), they also offer other benefits. At interchanges, travellers also usually need to wait and spend time. When services are offered, or shops are located at the interchange, the quality of the time spent waiting may improve. Comfort is nowadays a very important element of a trip, as well as the perception of security and reliability. This leads to several additional services at public transport interchanges besides the transport function itself. Many studies have identified important interchange elements for travellers, also making a distinction between the types of travellers. Commuters generally want to travel as quickly as possible, whereas leisure travellers tend to have fewer problems with longer waiting times at interchanges. Interchange accessibility and social exclusion aspects are important and as such have been identified as important elements within the City-HUB project.
- *Operators and managers of the interchange:* Of course, transport operators also use the interchange. Bus companies use the roads, bus lanes and other facilities at the interchange to bring and collect people. The same holds true for operators of metro, tram and train services. Usually (national) standards apply for the construction of physical infrastructure, depending also on the sizes of rolling stock. Interactive discussions will take place in the design phase between different stakeholders in order to construct an optimal interchange. An interchange may be operated and managed by different organisations. Railway stations in the Netherlands, for instance, are operated by Pro Rail, the company that is also responsible for the management of the rail network. But interchanges may be large and consist of different elements: a bus terminal, a rail terminal and access to the metro and tramways. For instance, in the Lille Europe station management, the SNCF and Transpole are responsible for station operations and bus public transport, respectively. In addition, regional transport authorities may also often be involved in interchange management and operation, with responsibilities for specific parts (e.g. urban public transport connections such as bus and tram stations). A joint venture of different stakeholders is often created to manage the entire interchange and coordinate processes. The operator of an interchange (the organisation that is responsible for the day-to-day management) aims, in the first place, to achieve complete satisfaction of the needs of the different users (the travellers and transport operators) and transport authorities (when

involved). The operator also needs to take the requirements of the (central or local) government into account: they may therefore face certain restrictions on activities. The legal framework and local circumstances will affect the daily operation of an urban transport interchange. However, generally the operator aims to manage a cost-effective terminal with the challenge being to find a balance between the different needs of users, operators and transport authorities (KITE 2008).

- *(Local) government:* Governments mainly use legislation to realise objectives. Interchange operators have a certain level of flexibility in how they operate; however, they need to obey the legal framework that is in place. In many countries, national or regional governments have developed a legal framework that sets standards, for instance, for the accessibility of certain groups to prevent social exclusion and improve safety. Spatial planning laws and environmental legislations are both in place and interchange developers and operators must obey these restrictions. Larger interchanges are nowadays often part of large redevelopment programmes within cities, which means that a wider range of legislation than only that relating to public transport applies. Different levels of government may be involved in interchange design and management. This can affect processes and decision-making. Multiple levels of government may be involved depending on the scope and type of the interchange. Besides national, regional or local governments (municipalities), regional transport authorities may also be involved in transport policymaking (in metropolitan areas such as Amsterdam, Transport for London [2009] in London and Consorcio Regional de Transportes de Madrid [2010] in the case of Madrid interchanges). In many cases, the impact of governments goes further since transport operators are partly or completely owned by public bodies.

- *Owners and businesses:* Of course, the aim of the owner is also to satisfy the needs of the different users of the transport facilities and to meet the legal requirements of the government, but since interchanges often include a wider variety of functionalities such as shops, offices, housing and so on, the focus of owners and business-related stakeholders is often much broader. From their perspective, commercial and business activities are also important, with the volume of the different users (the travellers and transport operators) as an important business driver. This latter importance depends on the type of interchange. In a fully integrated medium-sized building, for instance, the local impact in terms of overall economic impact (measured in jobs) is highest. In this type of interchange, it is not surprising that the economic and business interests should also be strongly represented in the governance process.

4.1.2.4 How to involve stakeholders:
stakeholder engagement strategy

People tend to respond better to a well-structured stakeholder process and a clear step-by-step process. Therefore, a stakeholder strategy should be developed and the following typical steps should be considered:

1. Specify the issues to be addressed.
2. Identify the stakeholders (groups and individuals, diversity and level of influence) to involve.
3. Set out the ways in which they are to be involved and their potential contribution.
4. Establish the consultation/involvement process and the options for giving input (including how you will treat input that is either out of context, beyond the limits of the project or not able to be considered for any other reason).
5. Execute consultation process(es): Care should be taken to ensure that all voices are heard and different groupings of stakeholders may be required. The level of impact and influence of the various stakeholders needs also be taken into consideration.
6. Evaluate and follow-up.

Suggestions from stakeholders may either be highly strategic (e.g. fare policy of public transport), narrowly focused (e.g. location of the bus stopping area), unfeasible and/or poorly rationalised. How the variety of ideas and suggestions will be evaluated and integrated (or not) into the final project should be set out early on in the process as well as how feedback will be given to those that will have taken part in the public consultations. While most people tend to focus on local problems, these problems are not always well articulated. Care needs to be taken, therefore, to ensure that there are enough channels or options for feedback open for those groups that may be less dominant or less capable of having their voices and opinions heard.

4.1.2.5 What is meant by public involvement?

Engaging stakeholders is an ongoing dynamic process. Managing a public consultation is usually carried out at times of major change. This may occur when an interchange is being planned, designed and built, or when it is being redesigned or upgraded. The approach required for each stage differs and this should be decided early on to ensure a successful outcome. The objective of stakeholder engagement is usually to build effective, trustful partnerships where decision-making around the

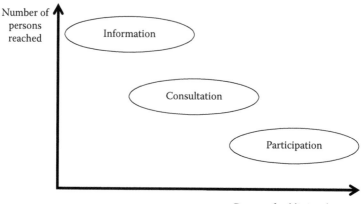

Figure 4.2 Different degrees of public involvement.

interchange has been carried out in a participatory way (sometimes called co-decision-making).

A broad distinction is made between involving the public and other interest groups (the 'citizen' input), and the public as 'travellers', where information is needed about their travel patterns, travel choices and so on. For example, travel surveys can include specific questions about the use of interchanges. Although on some occasions, the same technique may be used to collect both kinds of information, in terms of understanding the processes involved, it is better to consider the two functions separately. Figure 4.2 illustrates the different degrees of public participation.

Broadly speaking, three levels of involvement can be identified (Arnstein 1969):

- *Information*: The stakeholders are notified about the proposed implementation of an interchange plan. This is essentially a one-way process in which information is shared with stakeholders.
- *Consultation*: The views of stakeholders are sought at various stages of the formulation and implementation process, and are input into the process which remains under the control of the relevant professionals. There is a variety of ways that a consultation may take place, including via questionnaires, information days, leaflets and meetings. Listening and learning are part of consultation and should lead to a common understanding of the project. Consultation is more targeted at particular groups (e.g. travellers, operators, businesses, local residents, visitors), though it may involve large numbers of people in the structured process of information provision and feedback.

- *Participation (involving, collaborating and empowering)*: Stakeholders are involved in a two-way dialogue with the professionals, and have a direct influence on the outcome of the process; changes in attitudes and perceptions are likely to occur on both sides. Direct discussions between the various parties result in co-decision-making and the feedback loops of the process need to be carefully managed. Participatory techniques generally work best by involving far less people than is potentially the case with consultation or information procedures.

The distinction between consultation and participation is that the latter guarantees a stronger outside influence over the final outcome of the study, whereas with consultation, those in charge of the study have discretion as to whether and to what degree to take account of the results of the exercise. In general, the greater the level of involvement in the process, the longer the time and other resource requirements required during the development of the interchange. This greater influence should be offset by the improved quality of the decision-making process and the likely success of implementation. Success can be measured both in terms of a smoother administrative implementation (with a reduced number of formal objections), and by higher levels of public support after the scheme's introduction, both of which can reduce the costs and improve the delivery of the project. Many different public involvement processes exist and the selective application of different information, consultation and participation approaches can be found within the same study.

4.1.2.6 What kind of public to involve within the interchange decision process?

Decisions about who to involve in consultation or participation are crucial to the success of implementation in two respects:

- Firstly, if groups or individuals who feel that they are materially affected by the interchange are excluded, then this will cause resentment and increase problems in implementation (through public objections to the measures or failure to comply with any regulations), and may also result in a less effective plan than if the views of such people had been taken into consideration.
- Secondly, the range of people included in the process (and the way in which this is done) will have a major influence on the balance of opinion that this process uncovers. The spatial extent of coverage and the mix of people involved will have a major influence on the overall outcome of the process. (This is more of a problem with consultation, where people are only asked for their views, than participation, where people are encouraged to see different viewpoints and weigh these up.)

The involvement process should involve stakeholders within the catchment area of the interchange (possibly defined in terms of the journey to work area). The catchment area is also the appropriate unit for defining the relevant 'public' to become involved in the development of the interchange, particularly when carrying out consultation or inviting people to take part in participatory exercises. This geographical unit should contain the most regular users of the facility, including, if possible, those who do not live within the immediate built-up area itself. Where the local economy is strongly dependent on regional/national business links and on tourists, it will be important to take account of the views and needs of these groups as well. They can be consulted directly (e.g. via on-street or postal surveys) and/or their interests can be championed by local businesses who benefit from their trade. Finally, special attention should be paid to the question of how to engage with hard-to-reach groups (Tunbridge Wells Borough Council 2009), such as minority groups, people with disabilities (visual, cognitive, language understanding, etc.) and people who are not 'visible' in traditional engagement processes such as those depending on other helping structures.

4.1.2.7 Techniques for public involvement

The possible techniques and the strengths and weaknesses of each are reviewed under three stakeholder involvement levels:

1. *Information provision*: Different techniques can be used to invite people to take some action (e.g. attend a public meeting or complete an internet questionnaire) in order to become more involved in the consultation or participation aspects of the process. Information can also be provided by media coverage, posters, leaflets or the use of websites and participation may be encouraged by incentives. Information is also provided by exhibitions* that give an opportunity to set out in simple terms what is being proposed and invite feedback on particular issues.
2. *Consultation*: Public meetings that must be carefully prepared and well-argued represent the traditional means of public consultation, though they are often unwieldy, both in terms of the numbers of people who attend and the dynamics of the event. People with strongly held but unrepresentative views can dominate proceedings, and it may be difficult to gauge the true balance of opinion. Sometimes, these meeting can be targeted towards a restricted audience or specific groups giving more opportunity for in-depth discussions. Consultation can be organised by citizens' panels that are often large, demographically

---

* This may include a single stand in a public place (such as a shopping centre) or taking part in a more formally organised exhibition with other exhibitors.

representative groups of citizens used to assess public preferences and opinions. They are used to identify local issues and consult service users and nonusers. Citizens' panels can be used to assess service needs, identify local issues and determine the appropriateness of service developments.

3. *Participation*: Participation can be organised by the use of focus groups. These are a widely used form of public engagement, used to gauge attitudes and perceptions and to obtain input into the planning of interchanges and other transport features. The technique usually involves meetings with small groups of people (typically 8–10 people), and they are structured and executed in such a way as to engage with the audience and allow their opinions and suggestions to be heard. These processes may be called by another name: community x-change, consensus conferences, citizens' councils, deliberative focus groups or citizens' panels. The role of participants once a citizens' jury (in which members of the public can question and challenge the scheme's promoters, including their professional advisors) has taken place varies from nothing to being asked to help bring about the recommendations they have made (Figure 4.3).

Figure 4.3 Stakeholder participation. (Illustration by Divij Jhamb.)

4.1.3 Define functions and objectives of the interchange

The interactive engagement of stakeholders in the project and the processing of public participation outputs will provide a basis for defining which functions and objectives are expected for the new or refurbished interchange. This phase will consist of a list of expectations from each of the possible stakeholders and detail how they want to be engaged in providing their effective support to the project deployment.

Business expectations should be contrasted with public expectations to ensure that stakeholders' plans correspond to what users and neighbourhoods want.

4.2 VALIDATION

Proper mapping out of private sector interests (identification of stakes and stakeholders) in the interchange in various forms (investors, service providers, subcontractors, etc.) can effectively scale up resources available for interchange management from the time of the project preparation. On the other hand, cooperation between stakeholders means not only financial consequences but also consequences for common decision-making. Therefore, at the end of this step, the authority, as well as the involved stakeholders, should have a decision or an agreement on the chosen business model, including the governance structure.

4.2.1 Objectives of business models and financial plan

Business model analysis has the following objectives:

- Understanding the underlying principles of operating integrated systems (transport network/system, commercial parts, information systems, building/interchange infrastructure) which have multiple services and revenue obtained from various parts of the operation.
- The feasibility of various funding mechanisms and sources of revenue to be identified and utilised in a transparent and agreed manner needs to be assessed and best practices collected.
- Cost models, that is, determination of costs and benefits of solutions as in existing interchanges, for example, learning from existing cases and seeing how costs are distributed there. Using cost benefit analysis (CBA) will determine how much the services cost and which actors will cover which costs, including subventions. In the case of a lack of sufficient information on costs or benefits, other tools of financial analysis can be used. The main concern derived from the City-HUB

case and interviews is that interchange planning processes as they are today omit most financial and revenue stream analysis.

- Equally important to the cost model is the way revenues are pooled and redistributed in the integrated systems; for instance, in models where revenue collection is based on the joint efforts of interchange operators and private service providers. These can generate incentives for various partners to improve their operations and service levels in order to generate higher revenue. Equally, ways to reward or punish changes in efficiency and service delivery should be embedded in business models. Again, learning from existing models will allow for the consideration of the best functioning systems.

- Components of a functional business model should include clear methods for data utilisation and monitoring, and evaluation of the expected outcomes. These outcomes should cover all the above features and allow for the identification of possible areas for development in the case that one or more parts of the model are not functioning properly. In many cases, decisions made in the planning phase are not reviewed based on the actual data of interchange operations, including, for instance, the utilisation rate of rented properties, their rents compared to market rents in the surrounding area, changes in passengers flows and so on.

The business model should respond to a plan to cover the interchange costs. For any interchange, the basic cost structure consists of the investment, maintenance and management costs.

Investment costs consist of the initial construction of the interchange and additional refurbishment or upgrades, which impact on the balance sheet value of the interchange. These costs can and should be calculated over the entire operational life of the interchange, and charged through rents, user fees and so on if the investment is considered to be recovered. In the case of public sector investment decisions, there may be more altruistic considerations – providing for the public over the recovery of expenditure.

Maintenance costs are those costs that can be directly or indirectly attributed to the daily maintenance of the interchange. These include salaries of maintenance staff, equipment, cleaning, electricity and so on. Depending on the type of interchange, the size and role of these costs can vary significantly.

Management costs are those staff time costs that are consumed in the management of operations. Interchanges which have integrated shopping and other facilities are also likely to have full-time management staff, whereas non-manned stations are likely to need only a few hours of managerial working time.

If an interchange was operating on a profit or break-even basis, then the costs should remain below or equal to the benefits from operating the interchange. In this case, the revenues should cover a minimum of maintenance and management costs, and also investment costs if the interchange is considered to be financially self-sufficient.

On the benefits side, the following elements may or may not exist in the ownership and management model of an interchange: concession payments, rents of shops and other business entities, land rents, parking bay rents (buses), and parking charges.

The total monetary benefits will then depend on the possibilities which the interchange has for partnering with external stakeholders. In an optimal revenue generation model, the interchange is built on land which is owned by the transport authority, where additional land is rented for integrated shopping and business facilities. A concession model can be developed to run the interchange where additional income is collected from the various rents charged. Such a model requires careful financial planning in order to make a profit given the high investment costs required for integrated interchanges.

A minimum requirement for a future interchange following the City-HUB model is to have a financial plan covering the investment, maintenance and operating costs and the associated revenue flows, including all the relevant financial flows related to the operation of the interchange, whether or not those are public or private actors involved in the process. The framework for this should be developed at the planning stages, as it can pave the way for the consideration of the use of space, retail mix, rents, parking and so on. These considerations will also help the public authority to establish whether the proposed interchange could be feasibly managed by a public entity or if it makes more sense to have a public–private partnership through a concession or other type of arrangement.

Employment effects of interchanges depend on the services provided and the retail and other supporting functions which can be located at interchanges. Retail mix is the critical ingredient of a business model for an interchange. Retailers should be complementary to one another and also serve the vision of the attractiveness of the interchange. Retail mix allows the expansion of a secondary market around the interchange but can also provide a good markup for products available at the interchange. A good example of such services is a barber shop, which can attract passengers in transit with immediately available appointments, as opposed to a separate visit for such services. Of late, small shops catering for the supply of basic foods and other items required by busy urban professionals have gained importance at interchanges. In the development of public transport services, the service level and provision are critical determinants of the overall envelope of the service offering. However, so far, the way in which the provision of public transport services can be integrated into the surrounding economy and economic activity has been of little importance. This can be a potential way to significantly increase the utilisation of public transport services and a comfort factor for the passengers.

The different interchange types defined in Chapter 3 create various impacts on maintenance costs and associated revenue collection. The economic impact is low except in the case of fully integrated interchanges. Maintenance costs of cold/hot stations are low, and higher in the case of integrated models, due to, for instance, the provision of heating. The fully integrated interchange could follow a public or private ownership scheme. There is also the option of a joint venture.

4.2.2 Conducting analyses of interchange financial performance

In order to improve financial performance analyses of interchanges, those operated by the public sector need to be established as separate units for which costs and revenues can be accounted. This would offer better opportunities to monitor their performance and provide incentives to innovate private sector service provision in connection with the interchange. The process of estimating profitability can be further enhanced by the following steps:

- Form a business unit for the interchange. In the case of multiple partners, the cost accounting can be undertaken by the lead partner, such as the one bearing responsibility for the investment project. It is recommended that this is undertaken for all new interchanges already in the planning stages, so, for instance, maintenance models can be reviewed (in-house services or purchases from external service providers) in terms of their cost effectiveness; in later stages, the formation of business is more difficult in terms of taking into consideration all the sunk costs by partners up to that stage.
- Establish a clear revenue generation plan. There are opportunities to utilise the interchange area for private sector engagement in service provision, if rents are appropriate to the market potential; however, for revenue generation to be sufficient, the volume of businesses needs to be sufficiently large – one or two small shops will not pay for the maintenance of a large interchange.
- Integration with private sector investors and financiers. Partnering with the private sector to produce interchange services can be an effective way to generate additional resources and to speed up investment projects. For the private sector, incentives to do so would include the construction of shopping and business centres in high-density areas, where land is otherwise scarce. Furthermore, partnering with the private sector can bring in management knowledge of business models, which can complement the skills of the public sector; however, the financial profitability can still remain a challenge due to large areas requiring lighting, heating and maintenance.

4.2.3 Ownership and management

Ownership of the interchange is critical for the full cost calculation of an interchange; those interchanges where several operating entities provide co-management are usually not able to track the maintenance and operating costs as they are collected at different operator levels but not collectively by a single managing entity. Although whether the ownership is by a public or private entity as such is not directly relevant for the financial flow estimation, the case studies have shown that most profitable operating models include public ownership of assets, but concession agreements with private companies can generate the required management efficiency. The sell-off of assets by public sector actors can generate one-time revenues but they are not an answer to ongoing maintenance and operating cost recovery in the long run.

In principal, the following opportunities could exist for integrated systems:

1. Publicly owned and maintained interchanges where income from, for example, ticketing systems and retail provide the funding for interchange operations. This is not a feasible model given the subsidised nature of much public transport in Europe, so the true case is that interchange management costs are simply provided for out of the public authorities' budgets.
2. Privately owned and maintained interchanges, where, for example, ticket revenues and revenues from retailing are used to pay for the services rendered from the private service provider in exchange for interchange management. Some modifications of this model exist in the cases where, for instance, bus companies pay for the use of bay area slots, but these can also be publicly managed interchanges.
3. Public–private partnerships, where space and facilities are owned and operated under an agreed concession model, where revenues from public transport can be used to finance the operation of the interchange. In this model, the ownership of land and facilities will be a key determinant of the financial profitability of the operations for the transport operator; for instance, if the land is owned by the public authority, rents can be used to offset the rent of the interchange and maintenance of the interchange area, which could be under private ownership.

4.3 DEPLOYMENT

4.3.1 Manage/define roles and responsibilities

An Interchange Management Plan should be developed and signed up to which clearly states the roles and responsibilities of all those involved in the interchange through the different stages (i.e. design, planning, construction and operation). This is similar in approach to the Station Travel Plans, a partnership between rail operators, local authorities, bus operators and other stakeholders to improve access to stations (Rail Safety and Standards Board 2013). This should avoid fragmented decision-making and importantly ensure that the priorities of all of the different stakeholders are considered. It should also set out the priorities for future development of the interchange. Interchanges are dynamic facilities and the management plan should reflect this; it should be updated on a regular basis. Opportunities for those involved in the interchange (interested parties) to meet on a regular basis to discuss developments should be arranged – this allows issues to be discussed early on, safety to be regularly reviewed and other topics to be discussed, such as changes in service timetables and so on. It also allows the marketing and promotion of the interchange to be more coordinated –resulting in of increased footfall and throughput. One organisation could be put in charge of organising the agenda and timings of these meetings to ensure that they use the time effectively and are well attended. The governance model must take the economic dimensions of an interchange into consideration as well. It should therefore clearly define who bears liability and the financial responsibility for certain operations and what should be done in the case of significant revisions to the existing operational model. These could be caused by regulations on accessibility, emissions, ventilation requirements and so on. Collaborative efforts by all stakeholders would also be needed for serious upgrades and refurbishments. Again, a deeper understanding of the financial model is required in determining each stakeholder's role in these processes.

4.3.2 Urban economic impacts

The relationship between transport infrastructure investment and development has been studied for a long time. A review of this link is described by Simon (1996). The relationships between transport investments and their effects on the economic or social life of a city or even a neighbourhood are often questioned. More often, it is large-scale transport investments that are questioned regarding their effects on the national or regional economy. These transport investments are mainly highways, high-speed train lines or light rail projects. The direct causality is discussed and even rejected by Plassard (1977), and Offner (1993) has criticised the 'structuring effect' of transport investment. Nevertheless, we will see that if a transport investment such as an interchange is implemented jointly with an integrated development plan or an urban regeneration or redevelopment plan, it is more likely to induce economic and social impacts. Preston (2001) conducted a review of economic impacts due to transport investments. Nevertheless, the main transport investments usually observed are either highways or railways investments. At a microlevel, that is to say, an urban level, one of the economic impacts is bringing a skilled workforce closer to employment opportunities. The link between investments such as an interchange and economic development is difficult to observe; the causality of increases in land prices and the presence of the interchange is not clearly demonstrated. The question of the impact of a city transport node is rarely tackled at the infra-urban local scale. Nevertheless, the City-HUB project aimed to observe the possible economic and social changes that occur as a result of or after the development or redevelopment of a transport interchange. Effectively, the interchange must be examined under its three main functions (Richer 2008): the transport function, the city function and the service function. The urban function contains the urban services where the economic role can be highlighted by commercial activities, shopping mall construction, business start-ups and so on. Filion and Kramer (2012) have stressed the importance of realising transport investments at particular nodes linking the planning principles and transportation projects and land use proposals in order to achieve sustainability and smart growth objectives. The main reason for developing interchanges at these nodes is to reduce car dependency, improve accessibility and increase public transport use. It can also increase urban density, as was the case for the Kamppi Terminal area and King's Cross St. Pancras. Interchange development can also create completely new neighbourhoods, such as in the Lille intermodal transport node inside the Euralille development project (Figure 4.4).

This project analyses the services function and particularly the urban services associated with the transport interchange and the possibility of economic enhancement due to the presence of a city-hub. As highlighted by Banister and Berechman (2001), three conditions must be present in

Figure 4.4 Euralille, France. (Courtesy of Jan Spousta.)

order to induce economic development. Transport investment such as a new interchange must be of a significant size in order to provide new accessibility and new connections between transport modes. The economic context must reach a high-quality labour force and present underlying dynamic economic externalities at the local, regional or national level. The third condition is the existence of a political willingness to implement complementary policies in order to provide a better environment and to boost the transport investment to generate economic development. These complementary policies can involve implementing a transport hub as part of an overall larger integrated policy and/or a plan aimed at (re)developing links between economic activities and urban (re)development. It is also important to consider that the interchange may provide network effects given its role in the transport network at regional, national or international levels.

All the interchanges studied in the City-HUB project have local and regional roles. Out of 21 interchanges, 18 have national roles, linking main cities within the country, and 8 of the interchanges have international roles with long-distance international coaches: Mendez Alvaro South Bus Station in Madrid (Spain), Main Train Station in Den Bosh (Netherlands), Macedonia Coach Terminal in Thessaloniki (Greece), Moncloa in Madrid (Spain), Kamppi in Helsinki (Finland) with a link to St. Petersburg and Köbánya-Kispest in Budapest (Hungary) with a link to Romania. This international role can also be provided by international heavy rail links such as King's Cross St. Pancras Station in London (United Kingdom) and Lille Europe and Lille Flandres stations in Lille (France); with the High-Speed Eurostar trains, the latter should register network effects.

Multimodal city-hubs are more likely to impact on the local economy and land use when integrated policies (transport and urbanism) are implemented associated with policy makers' involvement (Di Ciommo 2004; Heddebaut and Palmer 2014); new neighbourhoods can be created and urban regeneration supported. New housing and new offices can be realised on top of or near to the city-hub as well as new shopping malls in or nearby the city-hub. Nevertheless, this economic and urban impact can also be observed if the city-hub is implemented or developed within an existing strong economic city centre. The direct link between the transport hub and economic development is not easy to establish. The size and the role of the city-hub within the transport network providing direct links with main cities such as national or regional capitals may explain the development around it. We have seen that the link between multimodal interchanges and their impacts on land use is not direct if there is not a strongly integrated development plan associated with policy makers' involvement (Banister and Berechman 2001). Some cities have jointly developed housing, commercial and business zones to develop their transport interchange. Cervero and Murakami (2009) suggest that it could be interesting to use the sale of land properties to develop the intermodal hub as has been done in Hong Kong. Further research could try to analyse the integrated land use plans offering new urban facilities – commercial, housing or businesses – and their related value due to the evolution of the transport city-hub. Investments can transform 'non-place' interchanges into new 'mobility environments' (i.e. a City-HUB), where people feel as if they were in a 'Mediterranean square' or a concert hall (van Hagen 2011). The main conclusions relevant for the City-HUB model are the identification of the three conditions for promoting economic development effects (size of interchange, reach high-quality labour force and political willingness to implement complementary policies). Moreover, an interchange needs to consider the environmental policy impact linked to construction and operation, especially the energy connected to the building's orientation, such as thermal insulation, lighting, passive heating and cooling strategies, automation, use of solar panels, low energy ventilation, ground coupling and solar water heating.

4.3.3 Interaction between interchanges and city

The stakeholders' consultation covered 21 different interchanges and identified their functionalities in the cities where they were located. Figures 4.5 and 4.6 show an analysis of how those interchanges interact with activities in the surrounding area.

Many interchanges described in the City-HUB project enhance accessibility and connectivity between transport modes (from 4 to 13 different transport modes), which is one of the three functions of city-hubs as described by Richer in 2008. These transport modes – walking to access and circulate into the interchange, cycling with cycling parking – are present in all of our samples except for Plaza de Castilla and Mendez Alvaro, the Macedonia Coach Terminal, Ilford Station and KETL Kifisou Interchange in Athens. Some offer cycle hire, such as King's Cross Station, the Main Train Station in Den Bosh and Leiden, or a free cycle service, as in Lille Europe and Lille Flandres with the V'Lille. Some interchanges provide links with the metro, such as Kamppi, Köbánya-Kispest, Moncloa, Plaza Castilla and Mendez Alvaro interchanges, King's Cross Station, Bekkestua Interchange and Lille Flandres and Lille Europe interchanges. In 2015, New Street Station will be connected to the metro as well as the Érd Intermodal Terminal, where underground works are under way, and the railway station in Thessaloniki is also planned to be linked to the metro. Three interchanges have links with tramways: Kamppi, Birmingham New Street Station and Lille Europe and Lille Flandres stations.

Most of the studied interchanges have links with heavy rail, except for Plaza de Castilla, Bekkestua in Baerum Municipality, Intercity Coaches at Magnesia Interchange in Volos, the Macedonia Coach Terminal in Thessaloniki, KTEL Kifisou in Athens and Moncloa Interchange.

Some interchange interviewees declared that some interchanges have not generated local economic development; for example, Macedonia Coach Terminal in Thessaloniki. Others say that they have not developed new

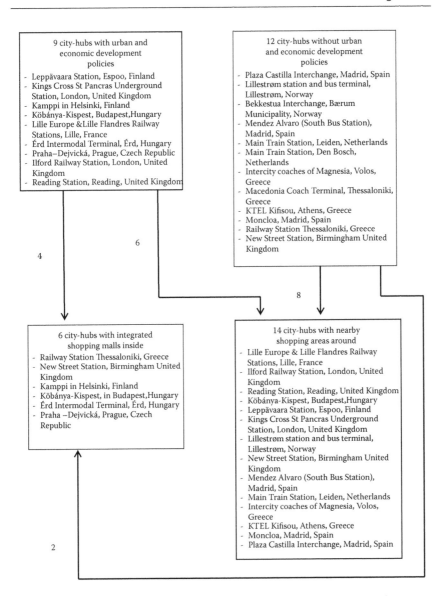

9 city-hubs with urban and
economic development
policies
- Leppävaara Station, Espoo, Finland
- Kings Cross St Pancras Underground
  Station, London, United Kingdom
- Kamppi in Helsinki, Finland
- Köbánya-Kispest, Budapest,Hungary
- Lille Europe &Lille Flandres Railway
  Stations, Lille, France
- Érd Intermodal Terminal, Érd, Hungary
- Praha–Dejvická, Prague, Czech Republic
- Ilford Railway Station, London, United
  Kingdom
- Reading Station, Reading, United Kingdom

12 city-hubs without urban
and economic development
policies
- Plaza Castilla Interchange, Madrid, Spain
- Lillestrøm station and bus terminal,
  Lillestrøm, Norway
- Bekkestua Interchange, Bærum
  Municipality, Norway
- Mendez Alvaro (South Bus Station),
  Madrid, Spain
- Main Train Station, Leiden, Netherlands
- Main Train Station, Den Bosch,
  Netherlands
- Intercity coaches of Magnesia, Volos,
  Greece
- Macedonia Coach Terminal, Thessaloniki,
  Greece
- KTEL Kifisou, Athens, Greece
- Moncloa, Madrid, Spain
- Railway Station Thessaloniki, Greece
- New Street Station, Birmingham United
  Kingdom

6 city-hubs with integrated
shopping malls inside
- Railway Station Thessaloniki, Greece
- New Street Station, Birmingham United
  Kingdom
- Kamppi in Helsinki, Finland
- Köbánya-Kispest, in Budapest,Hungary
- Érd Intermodal Terminal, Érd, Hungary
- Praha –Dejvická, Prague, Czech
  Republic

14 city-hubs with nearby
shopping areas around
- Lille Europe & Lille Flandres Railway
  Stations, Lille, France
- Ilford Railway Station, London, United
  Kingdom
- Reading Station, Reading, United Kingdom
- Köbánya-Kispest, Budapest,Hungary
- Leppävaara Station, Espoo, Finland
- Kings Cross St Pancras Underground
  Station, London, United Kingdom
- Lillestrøm station and bus terminal,
  Lillestrøm, Norway
- New Street Station, Birmingham United
  Kingdom
- Mendez Alvaro (South Bus Station),
  Madrid, Spain
- Main Train Station, Leiden, Netherlands
- Intercity coaches of Magnesia, Volos,
  Greece
- KTEL Kifisou, Athens, Greece
- Moncloa, Madrid, Spain
- Plaza Castilla Interchange, Madrid, Spain

Figure 4.5 Functionalities of the 21 interchanges analysed in the City-HUB project. (From Heddebaut and Palmer, In *TRA2014 Transport Research Arena 2014. Transport Solutions: from Research to Deployment – Innovate Mobility, Mobilise Innovation*, Paris, France, 14–17, 2014.)

housing or new offices around the interchange, such as Intercity Coaches of Magnesia in Volos, the KTEL Kifisou in Athens, the railway station in Thessaloniki and Moncloa Station in Madrid, even if the interchange pre-existed nearby shopping activities. The development around the interchanges described in the City-HUB project is the same as that described

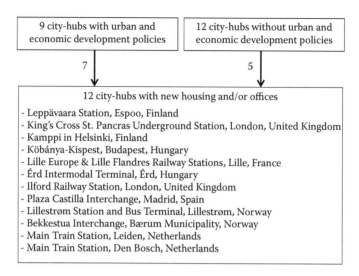

Figure 4.6 Impact on new housing and/or offices.

by Cervero et al. (1996); around rail (heavy or light) nodes, plans often concentrate offices, homes and shops to achieve more sustainable patterns of growth and travel. Cervero and Duncan (2002) and Ryan (1999) found that properties near heavy rail or light rail stations can raise land values; in addition, Cervero and Duncan (2002) found that living near services and public facilities such as restaurants, pubs and childcare centres also inflated commercial land values in Santa Clara County. In the interchanges examined within the City-HUB project, nine presented integrated development plans and strong political willingness that aimed to link together the transport function with local economic and urban development policies and/or urban regeneration policies (see Figure 4.5).

There are six city-hubs with shopping malls integrated inside the interchange (see Figure 4.5). These shopping malls were planned together with the city-hub development and built as a part of the city-hub. They can be seen as a direct economic impact of the transport interchange. City-hub plans implemented under integrated urban and economic development plans are more likely to include shopping malls inside the interchange. The integrated development plans also provide new housing and offices either on top of the city-hub, for instance in Kamppi in Helsinki, or in the area surrounding the city-hub by the realisation of urban regeneration or the creation of new neighbourhoods, like in Lille. In some cases, the property programmes linked to the realisation of a new high-speed railway station have increased the number of offices in the same way as the TGV Sud-Est stations in France, as studied by Mannone (1997). This is also the case for King's Cross/St. Pancras Underground Station in London, Köbánya-Kispest in Budapest and Lille Flandres and Lille Europe stations.

4.3.3.1 Impacts on jobs

The implementation of large interchanges is supposed to support a high level of employment, particularly if its realisation is combined with the settlement of a new shopping mall or business district. Nevertheless, very few direct jobs or associated indirect jobs were reported in our enquiries. An estimation is difficult, but for those that have responded, the number of direct or indirect jobs is variable. The direct jobs are counted from 20 to 350 persons for operating and maintaining the interchange and fulfilling the ticket-selling operations. This may increase if the drivers' teams (bus, metro, tram or railway) are added. There is a difficulty in attributing the indirect jobs to the city-hub development because there are no data available at the urban interchange scale. In the City-HUB sample, the indirect jobs generated ranged from 71 to 1000. In the case of Mendez Alvaro Bus Station in Madrid, 350 direct jobs and 1000 indirect jobs were announced, Reading Station estimated 1000 indirect jobs, as did Köbánya-Kispest in Budapest. In addition, the business practitioners interviewed for the Lille transport interchanges (Lille Flandres and Lille Europe railway stations) attributed all of the new jobs created (mainly in offices and commercial functions) in the new Euralille neighbourhood to the presence of the interchange; however, only a proportion of them are considered directly linked to the city-hub. The economic enhancement of the city-hub surroundings could be due to the central situation of the interchange and sometimes it could also have appeared without the existence of the interchange in cases where the neighbourhood is already economically attractive. In some cases, increases in employment in or around an interchange may be offset by reductions elsewhere, for example, in the retail sector. Moreover, the indirect jobs are not always linked to the implementation of the city-hub. In the case of the integrated development of the city-hub and new economic functions such as shopping malls, new offices and new urban functions, the links are more interrelated.

4.3.3.2 Coordination of land use and city integration

As described in the previous section on economic impact, when a global development plan exists, coordinating land use and city integration in order to integrate the city-hub construction within the urban context is likely to have an impact on housing and business activities around this interchange. When this impact is observed, it is possible in some cases to give an estimate of their costs. Sometimes, the development of a city-hub is an occasion to renew the attractiveness of a whole neighbourhood and the enhancement of the quality of the urban context may register high prices of housing, businesses and shops.

The most meaningful examples from the City-HUB project are King's Cross/St. Pancras Station in London, where a new neighbourhood has been constructed and/or regenerated; Ilford Railway Station in London, where residential and commercial activities were planned; the Kamppi Interchange in Helsinki, with housing built on top of it; and Leppävaara Station in Espoo, where new housing and businesses have been built around the interchange. The Lille Flandres and Lille Europe interchanges are located within the completely new neighbourhood of Euralille, which includes housing, businesses and a shopping mall that were developed at the same time as the Lille Europe railway interchange.

The case of Ilford Railway Station in London is interesting because there is not yet an observable impact on the surroundings of the new interchange, but it is integrated into a development plan that foresees the regeneration of the neighbourhood in early 2019. There has not really been any noticeable change as of yet. However, this may change in the future, following the opening of Crossrail (in 2019) and the other planned developments linked to this, including the regeneration plans within the Ilford Town Centre Area Action Plan and the Crossrail Corridor Area Action Plan. *That is to say, those impacts may occur long after the opening of the interchange.*

In Figure 4.6, we see that the creation of new housing and/or new offices is more likely to occur when there is an economic and urban development plan (for 77% of observed interchanges) than in the absence of this kind of plan (41%).

In order to measure the economic development around the transport interchanges, it could be interesting to establish a perimeter, for instance, a walkable area (500–800 m) around the city-hub, and to make an inventory of the buildings and their use (residential, commercial, mixed residential/commercial or institutional), as described by Diaz et al. (2012). When commercial activities are encountered, it is also important to note the nature of this commercial activity (product sale stores, services, restaurants, hotels, etc.). It may also be possible to observe the quality (low, medium, high) of these commercial activities and existing establishments. When it is possible, it may also be interesting to register the cost evolution of housing or buildings used for businesses and commerce. Enquiries could be used to identify the clients, such as travellers, interchange staff, residents, tourists and employees from other companies. Comparisons between the urban quality and specificity around the city-hub and the whole city area could then be made, as was done in the Diaz et al. (2012) study.

4.4 MONITORING AND ASSESSMENT

4.4.1 Start, implement and monitor performance

For monitoring and assessing the impact of the objectives of the interchanges, the following analytical framework can be applied. Within the framework, the extent to which the objectives (formulated in Phase 1 – Identification)

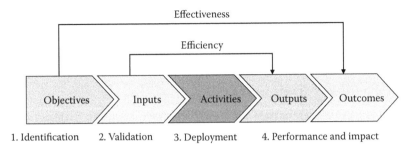

Figure 4.7 Monitoring and assessment in the City-HUB life cycle.

are actually realised can be assessed. The effectiveness of the objectives (are we aiming at and doing the right things?) are mostly defined in terms of economic spin-off and the number of jobs created. The efficiency (are we doing things right?) can be measured in terms of the supply of transport modes and services versus the actual use of this supply, for instance, in the volume of passengers and degree of utilisation.

In order to assess the effectiveness and efficiency of interchanges, SMART indicators need to be formulated for each of the objectives and inputs, implying that the indicators are: *specific* (detail exactly what needs to be done), *measurable* (achievement or progress can be measured), *achievable* (objective is accepted by those responsible for achieving it), *realistic* (objective is possible to attain) and *timed* (time period for achievements is clearly stated). For each type and each specific interchange, the objectives, inputs and corresponding indicators will differ depending on ambitions, characteristics, conditions, business models and so on.

In Figure 4.7, a connection is made with the four phases of the City-HUB life cycle. In Phase 1, the objectives are formulated. In Phase 2, the objectives are translated into inputs or validated plans. Phase 3 corresponds with the activities and in Phase 4 the impact is assessed and measured.

4.4.2 Medium-/long-term socioeconomic and environmental impacts

The local impacts correspond to the outcomes in the framework. These impacts are defined in terms of economic spin-offs for the local environment, such as the number of jobs created and commercial activity in terms of number of shops, offices and so on. It is recommended that these outcomes be measured at least once a year, as well as the extent to which the actual outcomes match with the expected outcomes that were incorporated in the original objectives and business model. Also, other soft impacts can be assessed, such as the contribution of the interchange to the image of the local environment and city. In addition, environmental impacts can also be considered important local impacts (outcomes). Lower emissions will contribute to a healthier local environment. It is recommended that clear and SMART objectives are set in the first phases for all of these types of impacts.

4.4.3 Feedback from users

Urban transport interchanges play a key role within urban transport networks since they allow different modes to be used in an integrated manner within the public transport chain. In this context, identifying and monitoring users' requirements is of particular importance to achieve the most appropriate policy measures for public transport interchanges, because they are particularly affected by the quality of service. Therefore, user perceptions should be considered as fundamental both in the planning process and operation management phases of an interchange. Moreover, user expectations should be seen as input in the evaluation framework, while the extent to which their expectations are met should be regarded as an important output. Users will be asked to provide feedback about the extent to which their expectations are met. This can be done at a later stage, but it is recommended that the relationship between the inputs and outputs is measured regularly in such a way that managerial or service adjustments can be made quickly.

4.4.4 Promote an innovative business model

The public sector has a tendency to apply the same logic and concepts over time. Innovations may not emerge from a cycle of repeated actions. It is therefore recommended that the public sector engages with private sector entities early in the planning of interchanges, as this may also bring out business potential that the public authorities have not mapped out. In some cases, such partnerships can also speed up the construction of an interchange by bringing capital into projects. This is another reason for early engagement. Investors will be able to set their own objectives at an early stage and make their own calculations regarding the economic profitability of the project. In the long term, such arrangements can provide support to the small and medium enterprises (SME) sector and provide local employment opportunities. As is the case for all stakeholders, actual outcomes need to be assessed and monitored regularly in order to make speedy managerial or other adjustments possible. These expected outcomes will often be formulated in terms of passenger volumes, the average time spent at the interchange and a multiplier for their general or specific expenditures during their stay in the interchange which are relevant for the investor.

In most cases, passenger volumes, time spent in the interchange and expenditures during that time are key determinants of what additional services and businesses can be offered in a profitable way. As such, accurate information on these figures is crucial business intelligence for these investors and their revenue generation or business models. Delivering such business intelligence regularly is therefore a key function for the owner and operator of the interchange. In the early stage, it will enable investors to make a solid business case; later on, such intelligence provides important management and monitoring information for these investors and also for other stakeholders. Since passenger volumes may differ between the various seasons, such intelligence should be delivered at least on a monthly basis in order to capture any seasonal patterns.

Part II

Elements of efficient interchanges

Chapter 5

Making a successful interchange in reality

Derek Palmer, Katie Millard, Clare Harmer,
Jan Spousta and Juho Kostiainen

CONTENTS

What makes an interchange successful? Various factors are important, including the integrated management of the interchange, integration with the local area, the relevance of different typologies (i.e. different interchanges need different facilities), essential and desirable features, and the different zones of an interchange.

Services are a key aspect in determining quality and success. These depend very much on the space available and passenger flows. It is important to consider the most appropriate use of space when deciding which facilities the interchange needs to contain. These will have implications for the functions provided and the services required. Figure 5.1 identifies three different zones within an interchange: the *Access/Egress Zone*, the *Facilities and Retail Zone* and the *Transport/Transfer Zone*, each with a different focus on services.

As shown in Figure 5.1, the Access/Egress Zone should provide facilities and services for the different types of users arriving at and leaving the interchange: pedestrians, cyclists and motorised transport (such as taxis or

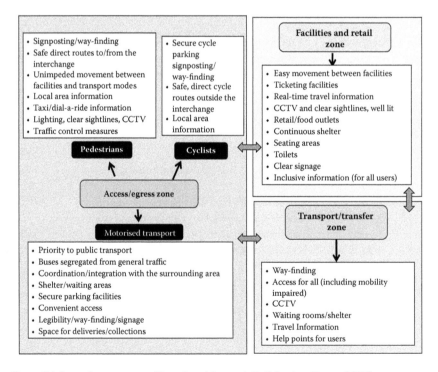

Figure 5.1 Interchange zones. (Based on Network Rail Station Zones 2011.)

kiss-and-ride). Key facilities that should be provided in this zone are those that assist safe, efficient movement in and out of the interchange, such as convenient access; signposting and way-finding; direct routes for pedestrians and cyclists with traffic control measures (such as pedestrian crossings where necessary); and information about the local area, including taxi and dial-a-ride information. For those with bicycles or vehicles, secure parking is essential, while waiting areas with shelters should be provided for those waiting for public transport modes.

The Transport/Transfer Zone is where users will be waiting for transport modes within the interchange. Here, there should be convenient, easy to navigate access for all. Waiting rooms and shelters fitted with CCTV for security should be available to travellers, with up-to-date travel information and help points if staff are unavailable.

The Facilities and Retail Zone is the part of the interchange where users who have more time available to spend (such as leisure travellers) can undertake activities such as shopping and eating while they wait for their transfer. Therefore, shops, food outlets, toilets and seating areas should all be provided here. This zone also covers ticketing facilities and should provide real-time information to ensure users are kept up to date with any delays or changes to their travel.

This chapter focuses on five sections: the first three correspond to the areas of the interchange; two additional sections focus on how to manage information services and intermodality; and the final section looks at to what extent all of these elements need to be considered according to different types of interchanges. The sections are as follows:

- Access/egress and the local area
- Transport and transfer
- Facilities and retail
- Travel information and intermodal services
- Scaling of services at the interchange

The Access/Egress Zone is where facilities concerning the local area and transport services are focused. Travel information and intermodal services are spread across both the Access/Egress Zone and the Transport/Transfer Zone. Facilities, including retailing, are based within the Facilities and Retail Zone. Some facilities are common, fundamental elements that are applicable to all types of interchange and should be considered within both the planning process and the operational management phases. In each of these zones, the facilities can be separated into those that are 'essential' and those that are 'desirable'. Essential facilities refer to those required to provide a basic level of service for users, with desirable facilities increasing the attractiveness of the interchange to users.

As is discussed later in the chapter, not all interchanges can provide the same level of service (often related to monetary or spatial constraints).

Small interchanges should aim to provide the essential minimum level of service, while larger facilities must provide this essential level of service with additional, more desirable facilities and services included. 'Desirable' facilities at a small interchange are increasingly considered as 'essential' facilities as the size of the interchange increases.

Several examples from the City-HUB project case studies and some other European interchanges are provided in this chapter in order to show the real implementation of the elements addressed throughout the different sections.

5.1 ACCESS/EGRESS AND THE LOCAL AREA

It is important to consider how the interchange fits into its surrounding environment, and design a structure with facilities that fit with local networks and destinations. By understanding how people access and make use of the interchange, it can be designed to have good links with external facilities, making it more appealing to users (Figure 5.2).

5.1.1 Local area facilities

In order to understand what facilities and services need to be provided, movements through the wider interchange area should be considered. The key things to look at are the local patterns of movement to and from the interchange by all modes and users and the key locations from where users come. This indicates the level of demand for different feeder modes and the facilities that will be required.

Figure 5.2 Access/egress and the local area: Stratford Interchange, London, United Kingdom. (Courtesy of Jan Spousta.)

The most common access mode is likely to be walking and cycling; therefore, these should be made priority modes in the local area, as well at the interchange. Pedestrian and cycle routes to the interchange should provide access to nearby facilities, such as shops, particularly if such facilities are not available within the interchange itself. Locating interchanges in close proximity to urban areas facilitates access by cycling and walking, and attracts greater use of public transport.

Providing multiple routes to and from the interchange ensures that users can easily access it and also helps to reduce travel times by providing more direct routes. Routes should be sized so that there is improved access both for those with reduced mobility and also to take account of differing pedestrian flows (e.g. during peak and non-peak times). Measures should be taken to avoid severance and barriers to ensure that the interchange is successfully integrated into the wider urban area.

Way-finding is essential in the area surrounding the interchange as it facilitates easy movement between the interchange and external destinations, along streets, footpaths, cycle routes and in public spaces. Signage should be clear, concise and easy to understand. Visual connections to the surrounding area are useful since this creates less confusion and uncertainty for users, and is particularly important for smaller street-based interchanges.

Safety is a key concern and therefore pedestrian and cycle routes should, where possible, avoid conflicts with vehicles. This can be achieved by locating parking facilities close to the interchange entrance/exit with clear signage and way-finding, thereby reducing the distance pedestrians are required to walk and minimising their potential conflict with vehicles. Where conflicts are unavoidable, traffic control measures should be implemented in order to increase pedestrian safety, such as low speed limits, speed bumps, cycle priority at entrances/exits and signal-controlled pedestrian crossings. Where the risk of conflict is high and where there is available space, physical segregation of pedestrians and vehicles should be introduced.

Improving the public realm, for example, through providing street furniture along routes or improving sight lines to external facilities, will make using the interchange more appealing.

5.1.2 Entrance/exit

Providing multiple access/egress points for different modes can reduce the likelihood of conflicts and increase access to and from the catchment area. Keeping routes as straight as possible and using symbols or markers for different modes also makes routes easier to navigate. If an interchange entrance/exit feeds into an external facility such as a shopping centre, it will be necessary to negotiate with the facility provider to allow for signage to the interchange to be displayed and maintained.

Figure 5.3 Entrance/exit, Paseo de Gracia, Barcelona, Spain. (Courtesy of Lluis Alegre.)

Security is a key consideration at the entrance/exit of the interchange as users can feel vulnerable here. Depending on the local context, and the surrounding area of the facility, a mix of CCTV cameras and human presence can be effective for waiting areas and parking and pedestrian/cycle facilities. Providing clear sight lines between the inside and outside of the interchange can increase users' perceived safety and security when there is an absence of staff. To ensure integration with the local area, information such as maps about the local area and routes to external facilities should be provided at the exit(s) of an interchange.

It is crucial for interchanges to provide effective access for people with reduced mobility or disabilities. Step-free access to and from the interchange is essential, which should include the use of ramps, lifts and escalators. It may be necessary to provide both step and step-free routes in order to maintain efficiency of movement at entrances/exits. Step-free routes should be provided in the most intuitive locations with long sight lines, and information on step- and obstacle-free routes should be provided, including braille maps, tactile/talking signs, audible directions or tactile paths (Figure 5.3).

5.1.3 Design and layout of access and egress modes

Different transport modes require different infrastructure and traffic management. This is reflected in the prioritisation of modes at the interchange as well as different terminal designs.

In general, pedestrian and cycling modes should have the highest priority, followed by public transport, including bus, train or metro, then access by taxi and kiss-and-ride (K&R), with personal car transport being the lowest priority.

The transport services will determine the infrastructure layout and traffic management priorities and customers' needs. The highest priority is usually given to long-distance and core network transport services, while infrequent complementary local services are allocated less space. It is important to consider this within both the design and operation of interchanges.

5.1.3.1 Rail-based transport: metro, underground, tramway and regional and suburban rail

Rail-based transport operates in a completely different way to road-based transport services. The points of interchange between the modes therefore need careful design. Access to metro and underground services generally takes place by walking. Direct, convenient and clearly identified routes should be provided to/from stations, with easy and unobstructed access.

Surface to underground transfer connections need to be carefully designed and it is important to remember the needs of passengers with reduced mobility. The regulations for the different modes also need to be addressed; for example, tickets may need to be revalidated when changing modes.

Specific rules apply to tram transport given that tram infrastructure is usually a part of the road layout with trams having right of way over other road users (including pedestrians). It is necessary to consider traffic safety and users' comfort. A very common means of addressing these are traffic light-controlled pedestrian crossings (also adapted to disabled users' needs) across tram tracks.

There is a significant difference between the operational needs of long-distance rail services (with particular attention to high-speed rail, which might require specific check-in procedures similar to airport check-in), regional and suburban trains and urban rail transport such as metro and trams.

The time and platform capacity required for long-distance trains is usually much greater which must be reflected in platform design. Long-distance passengers also demand additional non-transport services to be available at the terminal (e.g. waiting rooms, refreshment facilities, Wi-Fi, left luggage office and/or boxes). Additional layout requirements emerge where there is restricted access to the platform for non-passengers. In these cases, sufficient area should be available for people accompanying or welcoming travellers.

Suburban trains are commonly used by regular commuters who are familiar with boarding and disembarkation procedures and so do not spend much time at the terminal; the crucial task is efficient traffic management.

Metro and trams are high capacity with very high frequency which puts the emphasis on design and layout to support their punctuality and reliability as well as traffic safety for both vehicles and pedestrians.

Allocating each train its own regular platform which facilitates passengers' navigation and movement has proved effective. Trains due to travel in the same geographical direction should be grouped together in the same part of the terminal in order to avoid unnecessary movement of passengers.

Integration involving rail is more complex than integration between other means of public transport. Railway stations are often old, and are not necessarily easily adapted to 'access for all' standards and integration with other modes. Integration and coordination are also required between different stakeholders, who may have conflicting objectives. There can be a challenge whereby rail authorities do not want to collaborate with bus companies, as they consider buses as competitors, particularly long-distance buses.

5.1.3.2 Buses and coaches

Bus interchanges are more common than any other sort of interchange. Transfers may be relatively simple, such as a route change, or be more complicated, involving regional or coach services or a multimodal option.

There are several different bus interchange layouts. Different interchange locations will suit different designs; however, all designs should allow for easy access to and from bus stops that is clear of obstructions. Safety is also a key issue. Any furniture provided, for example, seating, should not impede movement, especially for those boarding and alighting buses. Good lighting is also essential in providing good natural surveillance and promoting personal security.

When designing bus facilities at an interchange, walking distances, convenience and legibility are key factors. Decisions on where to locate bus stands should be based on minimising walking distances to other services within the interchange. Organising bus stands logically makes them easy to understand, adding to their convenience. The simplest way to achieve this is through organising them geographically based on the direction of the travel corridors; on the other hand, it is quite common that longer distance bus and coach service stands are located in a different part of station to suburban services in order to avoid overloading by local passengers. Long-distance passengers usually have additional non-transport demands – they should be provided with waiting rooms with seating, refreshment facilities, Wi-Fi availability, left luggage offices and/or lockers.

Bus and coach routes should not conflict with other modes, including pedestrians. This is not only for safety but also to reduce potential delays to bus services caused by other traffic (Figure 5.4). There is a particular need to manage short-stay car parking, drop-off and taxi pick up effectively as these feeder modes often encroach into bus stop areas when not properly managed. Traffic control measures can prioritise bus movement within the interchange. When designing bus facilities, the operation of the surrounding road network and the movement and access needs of all users must be considered.

(a)

(b)

Figure 5.4 Passengers waiting for bus transport: Segregated from the moving traffic at Moncloa Interchange, Madrid, Spain (a) vs. not segregated at Macedonia Coach Terminal, Thessaloniki, Greece (b). (Courtesy of Jan Vasicek (a) and Jan Spousta (b).)

5.1.3.3 Cycling

There are two main focuses for cycling facilities:

- At the interchange, such as parking.
- Within the surrounding area, such as cycle routes.

Providing good cycling infrastructure in and around the facility can really help increase the attractiveness of using public transport. Undertaking a cycle route and facility audit, such as the Cycling Environment Review System (CERS),* will assess where improvements may be needed.

Cycle parking should be situated in a secure, sheltered location with good natural surveillance as well as closed circuit television (CCTV). It should be either very close to the main interchange entrance/exit, on the direct line of the cyclist approach or, alternatively, on the platform itself as long as there are no security concerns or problems with crowding, for example, at ticket barriers. Cycle parking should be easy to use, clearly signposted, require low maintenance and not impede pedestrian routes. Information on using the cycle facilities should be made available. Cycle parking availability should be based on current and potential demand. A count of existing numbers of parked cycles, including those parked 'informally' in the vicinity, is helpful for identifying the demand for additional cycle parking. In larger interchanges, cycle parking can be expanded into providing other services, such as maintenance. The parking and maintenance area can be sublet to the private sector or a local cycling non-governmental organisation (NGO).

To make cycle parking convenient, it should be located closer to the main interchange building than car parking and have display screens with service information. Parking should provide a minimum level of shelter so that cycle saddles are protected from rain and snow. Cycle parking should be free of charge wherever possible and for security purposes have the ability to lock both the frame and the wheel. The public usually accepts that more secure parking, such as locked boxes, should be paid for, but the prices should be kept as low as possible.

Cycle routes to interchanges should provide easy access to cycle parking. This requires them to be convenient, direct, safe and avoid any unnecessary detours or steps. There should be clearly signed routes to and from nearby residential and employment areas that are integrated with any wider cycle networks (subject to consultation with local authorities). Cycle routes should be comfortable, with good drainage and even surfaces and any barriers along the route, for example, related to traffic conditions, should be minimised.

Within the interchange, step-free access and automatic doors are recommended for cyclists wishing to take their bike with them on the rest of their journey. Lifts should be clearly visible and large enough to accommodate at least one full size bicycle and have enough space to cater for cyclists as well as other users.

Providing cycle hire/bike sharing facilities can be considered, particularly at larger interchanges or where the volume of cycle users is high. These facilities provide users without their own bicycle with another option of transport. They can be particularly successful at urban interchanges.

---

* For further information, see https://www.trlsoftware.co.uk/products/street_auditing/cers.

EXAMPLES OF CYCLING FACILITIES

Utrecht and Groningen, the Netherlands: Pilot bicycle parking system. Each cycle parking space was equipped with a monitoring device which relays a signal to a receiver and then a central computer indicating whether there is a cycle using it and how long it has been parked there. This allows passengers arriving at the station by cycle to see a screen displaying empty spaces, allowing them to find a space more easily.

Leeds, UK: Leeds Cycle Point, located at Leeds Station in Yorkshire, offers secure, fully staffed storage for over 300 cycles. It provides maintenance and repairs services on a 'bring in the morning, take in the evening basis', as well as bike and accessory sales, a cycle rental scheme, cycling information and demonstrations.

Guidance* on providing cycle facilities at stations is already available and this can be expanded for use at other types of interchange.

5.1.3.4 Walking

Walking for transfers between different modes within interchanges as well as access to the facility is important. Therefore, it is important that routes and facilities are of a high quality appropriate for the type of interchange. Assessing the quality of the pedestrian environment through the use of the Pedestrian Environment Review System (PERS)[†] or other street audits helps to identify the main axes and where improvements are required. In addition, undertaking a site audit of access routes and so on will help to identify where improvements are most needed.[‡]

Pedestrian movements should therefore be prioritised in the areas surrounding the interchange using traffic calming and pedestrian priority measures on the streets, such as signal crossings, level surfacing, speed tables and landscaping. Routes should be clearly signposted and designed for use by large numbers of pedestrians, with access to and from parking provisions. Park-and-ride (P&R) and K&R areas should also be well indicated and lit.

Efficient lighting, direct connections and clear sight lines are required whatever the size or type of interchange; however, larger interchanges should have high-quality paving, street furniture and lighting, whereas smaller interchanges will have more simple and basic design features. Comfort and

---

* The UK Association of Train Operating Companies has produced a 'Cycle-Rail Toolkit' available from www.cycle-rail.co.uk/links.
† For further information, see https://www.trlsoftware.co.uk/products/street_auditing/pers.
‡ One potential approach is described in Guidance on the Implementation of Station Travel Plans (Rail Safety and Standards Board 2013), and a template is available from www.stationtravelplans.com.

convenience of all types of pedestrian users should be considered. Universal design principles specifically help vulnerable users such as the elderly and those with disabilities; clutter and obstructions should be avoided to prevent navigation problems. All users, not only those with reduced mobility, will benefit from features such as dropped kerbs and tactile paving, and well-maintained landscaping, such as regularly trimmed trees and the removal of graffiti.

5.1.3.5 Car parking (park-and-ride – P&R) and kiss-and-ride (K&R)

Capacity for car parking spaces (P&R) and the balance between short stay- and long-stay needs to be considered during the design stage. These factors depend on the type of user and the local travel patterns. Location, charging rates and potential parking restrictions should also be understood at the design stage as these will impact on the demand for facilities. P&R facilities are commonly used by commuters to travel to terminals located outside the built-up area. Parking facilities should also be provided for motorcycles and scooters. These require less space to park; however, they are at greater risk of theft so they also need fixing points for locks or access to secure lockers.

K&R facilities are important for long-distance passengers who travel with heavier and larger luggage, and will therefore find it more difficult and less convenient to travel by public transport. K&R sites should be conveniently located without impacting on moving traffic. They should be well signed so that motorised traffic moves quickly and efficiently around drop-off points and it is made clear that the bays are not to be used for parking. They should not conflict with other modes such as buses, pedestrians, cyclists and taxis. A well-lit sheltered area should be provided for those waiting to be collected.

Provision of direct, continuous and safe pedestrian routes from car parking and K&R facilities into the interchange are essential and traffic calming and formal crossings should be provided to protect pedestrians. In general, provision for car parking needs to be integrated into the interchange design but a higher priority should be given first to access by walking and cycling, and public transport. This can be modified if the interchange is designed as a major intermodal hub. The majority of trips should come from public transport so the interchange serves as a point of transfer. However, K&R can be an important feeder mode so should be fully catered for. Opportunities for alternative forms of car use for access/egress should also be encouraged where possible, for example, through provision of a dedicated car club or car-sharing spaces, or electric vehicle charging points. This may require negotiations with third parties. Where implemented, relevant information should be readily available to users within the interchange.

5.1.3.6 Taxis and community transport

Taxis are an important 'last mile' connection of a journey. Taxi ranks and mini cab offices require safe access, good lighting and clear signage and should be located in areas that are convenient to users. Locations of taxi ranks should be made clear to users of other modes through additional signage and road markings; users of private vehicles should be discouraged from using these areas and instead use K&R facilities. Locations for collection and setting down should be clear and support the layout of other transport modes at the interchange, and not impede local traffic flow.

To maximise user convenience, taxi ranks should be positioned so that the passenger door always faces the kerbside and waiting areas are sheltered from the weather. At interchanges where a dedicated taxi rank is not feasible, information about local taxi companies and a telephone facility should be readily available.

Dial-a-ride and community transport facilities should be available for those with limited mobility and located in areas that are safe and minimise conflict with other movements. Their location must be convenient in order to avoid vehicle drop-off taking place in more convenient, but unofficial, areas.

A summary of 'essential' and 'desirable' features for transport services is shown in Table 5.1.

Table 5.1 Essential and desirable features for transport services

Users/mode	Essential	Desirable
Pedestrians	Safe direct routes to/from the interchange Unimpeded movement between facilities and transport modes Signposting/way-finding Local area information and maps Lighting, clear sight lines, CCTV Taxi/dial-a-ride information alongside telephone access Traffic control measures, such as pedestrian crossings	Street furniture, landscaping, not an impediment to movement Segregation from traffic Easy access/egress to and from the interchange
Cyclists	Secure cycle parking (sheltered, CCTV) Safe, direct cycle routes to/from the interchange Signposting/way-finding Local area information and maps	Street furniture, landscaping Segregation from traffic Easy access/egress to and from the interchange
Motorised transport	Priority to public transport movements Co-ordination/integration with surrounding transport networks Shelter/waiting areas for buses Secure parking facilities Way-finding/signage Convenient access to P&R/K&R and taxi facilities Local information Space for deliveries and collections Short distance between car parking and the interchange to minimise conflict with pedestrians	Buses segregated from general traffic Street furniture, landscaping

5.1.4 Non-travellers

Interchange users can include non-travellers, so it is important to understand the movement flows of local residents and through traffic in the surrounding area. Having a clear passage zone allowing free access through the interchange without having a ticket encourages people to make use of interchange facilities even if they are not travelling. This is especially attractive for larger interchanges that offer retail and commercial options.

5.2 TRANSPORT AND TRANSFER

As previously mentioned, the Transport/Transfer Zone is where users are moving between and waiting for transport modes within the interchange. Not only are the facilities provided in this zone important but also the design and layout of the interchange is important to enable easy navigation and quick transfers. This helps make multimodal transport more attractive to travellers.

5.2.1 Design and integration of transport services

As part of the City-HUB project, a number of key factors were identified that can contribute to a successful interchange design.

Travel and time associated with distances between transport services are important to travellers as each transfer can accumulate substantial time within a door-to-door journey. It should therefore be a goal of an interchange to have the distances between transport services as short as possible. Not only is the distance important, but the connectivity between modes is another feature which promotes multimodal trips. In general, there are shorter distances between the same modes of transport (e.g. bus to bus, rail to rail); therefore, distances between different modes should look to emulate similar distances where possible (Figure 5.5).

EXAMPLES OF DESIGN AND INTEGRATION OF TRANSPORT SERVICES

Moncloa Interchange, Madrid, Spain: Short metro–walking and bus–metro distances.

Vauxhall Cross Transport Interchange, London, UK: The recent redevelopment of this bus interchange has specifically focused on the proximity and integration of bus, rail and underground services.

Figure 5.5 Connectivity and integration of transport services. (Illustration by Divij Jhamb.)

Capacity, open space and logical passenger movement are also important. Interchanges need to be designed so that they provide logical and easy passenger movement. Overcrowded areas and long queues to get through ticket barriers reduce traveller comfort and efficiency. A poor-quality travel experience is one of the key reasons given for not choosing to travel by public transport.

Coordination between transport modes and waiting are crucial for passengers. Public transport attractiveness is closely linked to the relative travel time compared to that of cars, making transfer time an important element of journeys. Waiting times of over ten minutes can make public transport less attractive to travellers. Integration of the timetables of different modes is a key way to reduce waiting times and can be more effective and less costly than increasing the frequency of services. This can be a challenge where different stakeholders are involved.

5.3 FACILITIES AND RETAILING

It is important to consider the facilities and services carefully in the design and operation of an interchange, as they can have a significant impact on how it is used. Benches and cafés, for example, encourage people to consider an interchange as more of a meeting place than a transport hub.

Figure 5.6 Photograph from inside King's Cross St. Pancras Station, London, United Kingdom. (Courtesy of Jan Spousta.)

The characteristics of an interchange have a very strong influence over the type of facilities and retail options available. There can be physical limitations, for example, in the redevelopment of existing interchanges, with the available site and buildings affecting the design and the requirement to maintain operational and functional capabilities while it is undergoing construction works.

The range of facilities at an interchange is also dependent on the type of interchange, with a very simple one requiring basic facilities such as shelter, seating and vending machines, while larger interchanges with higher service levels will also require toilets, kiosks and information desks. For new central interchanges that can combine transport and shopping centre functions, high-quality facilities and a wide range of retail options should be available (Figure 5.6).

The external environment surrounding the interchange also has an impact on potential retail opportunities as it is important to complement and support the local economic activities rather than necessarily compete with them.

5.3.1 Organisation of facilities

Facilities within interchanges need to be well organised to ensure they can operate efficiently. An interchange can be split into three distinct areas: the decision space, the movement space and the opportunity space. Each has different design requirements and functions, particularly in terms of

Figure 5.7 Platform for metro at Bank Interchange, London, United Kingdom. (Courtesy of Jan Spousta.)

facilities, and it is recommended that efforts are taken to ensure that these different areas do not conflict or interfere with one another. Spatial organisation determines how users move around and should be complemented by way-finding and information displays. The types of facilities needed by different users, both arriving and departing, as well as users with special needs, should be considered.

When planning the location of different facilities within the interchange, decisions should be informed by careful modelling of crowd density and crowd flows, particularly in larger facilities. Facilities have the potential to increase crowd density and reduce crowd flows; therefore, it is important not to locate several facilities within a movement space, for example (Figure 5.7).

5.3.2 Essential facilities for all types of interchange

The following facilities are essential in the design of a smart, efficient interchange. However, the scale of these facilities depends on the characteristics of the interchange and its users.

5.3.2.1 Service areas

These are the areas where services need to be provided, for example, ticket booths, information desks and first aid stations. Users will want to minimise the time it takes for them to purchase tickets; thus, adequate ticket booths should be open, depending on the anticipated customer arrivals. It should also be possible to purchase tickets from machines and online prior to arriving at the interchange.

Information desks should be placed at key points, be easily visible and accessible by the users, with long queues avoided. They should be located where no conflicts with pedestrian flows will be created.

First aid stations should be available at larger interchanges and managed by well-trained personnel. Users should be able to locate this station upon arrival at the interchange.

5.3.2.2 Waiting areas/platforms

These are essential to ensure that users feel comfortable and secure while waiting in an interchange. The quality of waiting areas is important in making an interchange successful.

Waiting areas, with comfortable seating and areas for standing and leaning, should be provided. They should be located in areas close to facilities and services and on transfer routes, close to transfer information. Waiting rooms should be well lit, heated and/or ventilated. Benches, for example, should have arm rests or be grouped in such a way that they cannot be used for sleeping. Appropriate levels of seating should be provided, while avoiding closely spaced seats that minimise user comfort. Passenger information should be provided both on screens and through audio announcements. Areas should make maximum use of natural surveillance, as well as being monitored by staff and CCTV. Waiting rooms should also be cleaned and maintained regularly to ensure comfort and attractiveness.

Passengers need to be protected from bad weather, that is, heavy rain, wind, snow and so on. Continuous shelter should therefore be provided throughout the interchange and over any outside transfer areas (Figure 5.8).

Restrooms should be provided for all users, and in convenient locations. Accessible toilets should be provided that are equipped for those users with special needs, for example, by providing enough space to manoeuvre

Figure 5.8 Train platform at Lille. (Courtesy of Jan Vasicek.)

wheelchairs, and making use of handrails and door handles that are easy to grasp. Baby care facilities should provide space for baby carers to feed and change babies and small children.

5.3.3 Essential features to consider in the design of an interchange and the facilities provided

5.3.3.1 Safety and security

Safety and security is highly important. There are usually two major aspects to safety and security – one is rational and the other is perceived (or emotionally triggered). In terms of the rational provision of safety, the facility must comply with regulations and provide adequate precautions, such as hand rails and anti-slip flooring and so on, to prevent any accidents. Areas of conflict between pedestrians and vehicles should be minimised and the use of traffic control measures put in place. Emergency exits should be clearly indicated to users and staff should be fully trained in emergency response.

Providing adequate levels of security is more challenging, as this can be subjective. Passengers usually respond best to human interactions, and the visible presence of people clearly recognised as providing this service is most effective. Depending on the local context and the surrounding area, a mix of technical, including CCTV cameras (especially for the outside areas such as outside bus stops, parking and cycle facilities), and human presence is usually the most effective. Security services can also be brought in from outside (under contract) in larger interchanges.

5.3.3.2 Accessibility

Making transport systems accessible for all people is an important part of achieving an inclusive society. In many countries it is also a high political priority to design transport systems for all. An accessible transport system will benefit all user groups.

Interchanges must provide effective access for people with reduced mobility or disabilities which will in turn benefit the majority of travellers including the elderly, people with small children, people with temporary accident injuries and so on (GUIDE 2000). In this way, social inclusion and mobility issues at the interchange are enhanced, creating a friendly environment for all users.

Ideally, step-free access should be provided between all parts of an interchange, allowing all travellers to use the same routes. Interchanges should be designed with the minimum number of levels possible, and where level change cannot be avoided, lifts and escalators as well as steps should be provided. Step-free routes should not be isolated from the main passenger routes and should be provided in locations with long sight lines. Information on step- and obstacle-free routes should be provided by proper signs, including braille maps, tactile/talking sings, audible directions or tactile paths.

EXAMPLES OF ESSENTIAL FEATURES TO CONSIDER IN THE DESIGN OF AN INTERCHANGE AND THE FACILITIES PROVIDED

Birkenhead Bus Station, UK: This was designed with passenger security in mind and therefore much of the station's structure is made from large panels of clear, toughened glass, giving clear sight lines to the surrounding area.

Lan van Noi, The Hague, the Netherlands: As the light rail and metro make shared use of platforms, level boarding is provided through the use of ramps that change the height of the platforms.

Bebra, Germany: There are sliding ramps for suitcases/bags to help improve accessibility for those with luggage.

Lille, France: A door to door service for the mobility impaired called Handipole is operated by Transpole for the LMCU. This is for the Lille Métropole residents. Within Lille Europe there is a welcome help desk for the mobility impaired.

King's Cross St. Pancras, London, UK: Fully accessible step-free routes from street to platforms are available although some routes are convoluted because of the legacy layout.

When designing for lifts and escalators, their location should be close to movement spaces or desire lines in order to improve trip or transfer time, minimise the chance of conflicting passenger flows and increase security by avoiding isolated areas. Natural surveillance helps to improve passenger security when using lifts. Waiting areas at lift entrances should be assessed in order to avoid any conflicts with pedestrian movements in adjacent areas. Lift dimensions should be suitable for both wheelchair users and those with luggage or pushchairs. Access to lifts and escalators should be clear of any obstacles.

Staff should be available to assist all users, including those with special mobility needs, such as helping them to purchase tickets, access platforms and other facilities. Mobility assistance buggies could help mobility-impaired passengers to transfer between modes and other facilities.

An interchange travel plan can be useful in order to assess accessibility. This is a management tool for improving access to and from a station and mitigating local transport and parking problems.

5.3.4 Desirable facilities for interchanges

Desirable facilities may be seen as essential to large central interchanges with high volumes of passengers and services, but desirable for smaller ones, with limited funding and available space.

5.3.4.1 Amenities

Although amenities may not be considered as necessary for users, they improve passengers' experience during their time in the interchange. Retail outlets allow waiting time to be usefully used. Coffee shops, restaurants, entertainment, play areas, pharmacies, tourist information desks, bank branches or automated teller machines (ATMs), post boxes and even accommodation are all useful. For larger interchanges, the choice of amenities helps to make the facility a destination for customers, rather than just a place to transfer.

5.3.4.2 Internet access

The provision of internet access is increasingly important following technological development over the last decade. It enables travellers to use the internet, browse, check their emails and so on, and for this reason, interchanges should offer internet access for those who wish to use their laptops, tablets or phones for work or for pleasure while waiting for their connections. This is particularly useful at interchanges with high volumes of commuters. Smaller interchanges with fewer services, and therefore longer waiting times, may also benefit as this will add to the comfort of passengers.

5.3.4.3 Comfort

Although comfort is subjective, it is closely related to the availability of facilities in an interchange and makes the time users spend there more pleasant. There are various issues to consider relating to comfort, including the use of space and seating arrangements, temperature, noise levels, security, cleanliness, lighting, access to amenities and so on. For example, it is important that background noise is low to ensure that any audio announcements can be heard.

5.3.5 Retailing

Facilities such as retail outlets should strike a balance between making enough money to remain viable and creating a vibrant environment while not causing conflicts in movement areas. Care should be taken to avoid operations by retail outlets within movement spaces, as this obstructs sight lines and movement, leading to confusion for users. When deciding which facilities and amenities to provide within an interchange, consideration of the size, location and level of service should be taken.

Retail services available nearby affect business opportunities for shops in new interchanges; therefore, it is useful to consider local demand when deciding on the location of shops. As well as providing direct employment,

(a)

(b)

Figure 5.9 Photographs from inside St. Pancras Station, London (a) and Köbanya-Kispest Interchange, Budapest (b) highlighting the different retailing within an interchange. (Courtesy of Jan Spousta (a) and Andres Garcia (b).)

expansion or redevelopment of an interchange can be a catalyst for regeneration of the surrounding area. Retail outlets should be complementary to one another and also improve the attractiveness of the interchange (Figure 5.9).

Particularly with retailing, planning is needed in the design stage to provide for the delivery of goods and materials and the removal of waste so that these operations have minimal impact on users or the day-to-day running of the interchange. Operators/retailers should seek to coordinate their delivery and removal needs to minimise the number of collections required and increase efficiency. Appropriate access routes and loading facilities need to be designed in larger urban interchanges, for example, using a central delivery point or zone where all deliveries are made and from where each recipient can collect and deliver.

EXAMPLES OF RETAILING AND FACILITIES

Shopping facilities at the interchange: Kamppi Interchange is in a very central location in downtown Helsinki. The interchange is combined with a shopping centre with an area of approximately four hectares. In addition to the shopping centre there are also offices and flats in the same building complex. A total of 170 businesses operate in the Kamppi shopping centre, including 106 stores, 35 restaurants and cafés, and 29 services such as beauty salons, gyms, banks and laundry. Köbánya-Kispest Interchange in Budapest also has a shopping mall adjacent to the interchange.

Clear strategy for retailer quality: In Moncloa, Madrid, the concessionaire authorises which services and activities are allowed to take place within the interchange (securing a distribution of types of businesses). It is up to the concessionaire to set the standards for the services provided, for example, opening times, comfort, and distribution of merchandise.

Surveys for monitoring the level of service: In Helsinki, regular common surveys concerning the level of service for the whole Regional and Local Public Transport are conducted. Some special studies have also been made on interchange safety and security issues.

High customer satisfaction was the reason for success at King's Cross St. Pancras Underground Station, London, and the Main Train Station in Leiden.

Waiting rooms, protection against weather, Wi-Fi access: At Moncloa, Madrid, Kamppi, Helsinki, and Köbánya-Kispest, Budapest, all transfers are completed inside the interchange building. Consequently, they have protection from the weather, as well as having short distances to the various services provided. Moncloa, Kamppi and Ilford, London, all provide Wi-Fi access to interchange users.

5.4 TRAVEL INFORMATION AND INTERMODAL SERVICES

The integration of information systems and ticketing, as well as the introduction of other Intelligent Transport Systems and Services (ITS), are essential for improving the experience and appeal of using public transport at interchanges. Clear, accurate and timely information helps reduce stress and improves the efficient use of interchanges. Information benefits walking and cycling access and egress as well as the smooth handling of emergency situations.

For passengers, the most important information is about the trip and interchange. The ease of transferring between modes and awareness of when and where transport is leaving is the crucial information supporting the main functionality of the interchange.

In addition to providing information to travellers, intelligent systems can measure and monitor traffic and people flow and therefore provide valuable information for design decisions and improvements to the interchange and public transportation services.

Depending on the characteristics of the interchange, information needs vary. In non-English-speaking countries, or where there are large numbers of overseas visitors, providing multilingual information is important. Accessibility concerns are relevant for all types of interchanges, but the solutions differ from tactile surfaces and ramps to elevators and auditory information.

5.4.1 Coordination of intermodal transport services

An integrated ticketing system reduces the need to buy tickets each time travellers change their mode of transport. The capacity and location of ticket offices becomes of greater importance without such a system. Short distances between modes can be offset by the inefficient location and lack of capacity at ticket offices, which increases the chance of missing transfers and leads to increased journey time. The regulations for the different modes need to be addressed and it can be the case that tickets need to be revalidated when changing between modes, reducing the efficiency and appeal of modal change.

Smart ticketing is at the core of interchange functions. For infrequent travellers, or people who aren't local, it is essential that the ticket options (such as zones and validity) are clear and purchasing the tickets is easy. Integrated ticketing makes the use of different modes and services from different operators easier for the traveller, thus making public transport more attractive. An overarching body that can handle fare structures may be a solution for creating more widely integrated ticketing systems that need the cooperation of regional authorities and stakeholders. The future of intermodal trips and the use of interchanges will depend on the introduction of smart card systems. Future smart card systems should also include parking fees for private transport (i.e. bicycle or car).

5.4.2 Travel and transfer information

Seamless intermodal journeys require integrated information. Interchanges therefore need to provide the information for different operators and for different modes clearly, yet comprehensively. Traveller information needs to be easy to understand, accurate and accessible where the passenger needs it. This should include pre-journey information; in-journey information; and real-time information.

The accessibility, availability and clarity of information are important (Figure 5.10). Information for different operators and modes should be presented in a harmonised manner and integrated to provide comprehensive

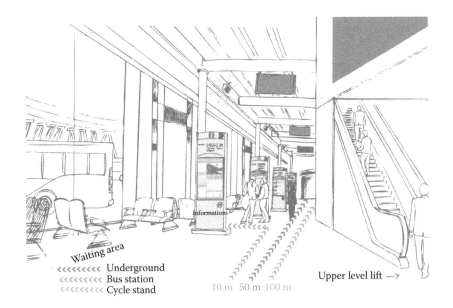

Waiting area
<<<<<<<< Underground
<<<<<<<< Bus station
<<<<<<<< Cycle stand

10 m   50 m   100 m

Upper level lift →

Information

Figure 5.10 Travel and transfer information. (Illustration by Divij Jhamb.)

EXAMPLE FOR TRAVEL AND TRANSFER INFORMATION

Helsinki Regional Transport Authority, Finland: The journey planner provided is very advanced and user-friendly and provides multimodal guidance (including walking/biking), saving preferences (e.g. route with fastest or least transfers) and the ability to pick locations from a map. The service has free Application Programming Interfaces (APIs) for anyone to develop journey planner applications using the timetable (and real-time) data for the public transportation and routes. For long-distance travel, there is a national journey planner available as well.

information services both at the interchange as well as for online services such as multimodal trip planners. The role of information services on the internet is significant and information regarding connections and transfers should be easily available from one source. In addition, personalised information can be offered to provide information that incorporates the needs of different users and includes contextual or preferential options. Near field communication (NFC) is an example of technologies that can be used for this type of interaction, including ticketing.

Typical information needs include departure, schedule and route information, which should be provided both as real-time information at the interchange as well as online. Information that affects travellers' plans significantly, such as interruptions or delays, should be provided clearly and timely along with alternative options.

5.4.3 Way-finding

Providing people with navigation aids assists the transfer of routes or modes and travelling to and from the local vicinity. Ideally, an interchange should have a self-explanatory design that is simple to follow and thus minimises the number of signs required. However, to ensure maximum usability, way-finding signs can be provided to complement the interchange layout. Clear pathways inside the interchange should be easy to identify and remember.

Way-finding is very relevant for cities as they tend to have high volumes of new travellers (e.g. through tourism) or less frequent public transport users. Simple and clear signing within the interchange helps travellers transfer whereas confusing signing might lead to increased travel time and passenger frustration. This is important for travellers new to the interchange, who, in particular, rely on signs for guidance.

Information can cover a range of subjects other than simply transport services, such as identifying different local facilities, temporary information, and safety and warning information. Each of these information types should be visible and clear within all areas of the interchange facility. Information can be displayed in electronic form, static signage and also be available from staff. There is a full range of media available to use including audio, visual and tactile to meet with the needs of all users. The information system should be planned as part of the initial design or upgrade (Figure 5.11).

Preparing a way-finding plan at the early stages of design is advisable. Key elements include the design of legible, well laid-out spaces; locating signing and information when and where users require it, for example, at key decision points like entrances and exits; using surface treatments, materials and lighting to complement signing; and making use of public art or landscaping features as landmarks to help define pathways to and from the interchange and help individual mental maps, including special needs groups. A balance is needed between the provision of way-finding signs, retail frontages, advertising and so on, and coordination should be ensured in order to avoid clutter and visual conflicts.

Way-finding can also be supported through various technologies, such as audio and visual displays, mobile telephones, and Bluetooth devices. The design of signs should include consideration of cultural and language differences, cognitive impairments, visual impairments and mobility impairments.

Way-finding should also be provided in the area surrounding the interchange. This facilitates easy movement between the interchange and external destinations, along streets, footpaths, cycle routes and in public spaces. This may require adopting a third party way-finding system in neighbouring areas.

Figure 5.11 Way-finding and facilities. (Illustration by Divij Jhamb.)

Key elements that influence legibility and should be considered in an interchange are: (i) layout, (ii) lighting, (iii) surfaces and materials and (iv) finishes and furniture.

Appropriate lighting should provide easy definition of routes and destinations in order to support users' navigation. Illuminated routes should be evenly lit, and any sudden changes in lighting levels, glare, dark spots or pooling should be avoided, in order to minimise the confusion of visually impaired users.

Travellers should be able to easily move through the space within the interchange to reach their destination. Multiple routes should be offered to and from the interchange thereby helping to reduce travel times, and routes should be sized so as to improve access throughout the interchange, letting passengers move in straight lines, and not on diverted pathways.

EXAMPLE OF WAY-FINDING

Moncloa Interchange, Madrid, Spain: This interchange has over 287,000 travellers each day; however, due to its design, it is still easy to move around and make connections as the interchange is never very overcrowded. The interchange is made up of four different levels without many physical barriers. A clear colour strategy, with consistent signage and symbols, also contributes to logical passenger movements and enables people to easily orientate themselves.

5.5 SCALING OF SERVICES AT THE INTERCHANGE

In Chapter 3, we settled on a typology of interchanges according to their two dimensions: (1) functions and logistics, and (2) local constraints. The first dimension depends on demand, modes and services and the second dimension on location in the city fabric.

It is useful to look at interchanges in terms of the different typologies as they can have different levels of demand, different modes of transport and different locations which in turn affect the number of stakeholders and integration with the local area, as well as the local impacts of the interchange.

As a consequence, size, functions and location of the interchange are all important when deciding upon the services and facilities to provide. It is important for transport operators that the quality of facilities is appropriate for the type and design of the interchange, in order to ensure that it is economically viable. It is also important to provide sufficient facilities to make the interchange appealing to users.

Three main typologies of interchange are focused upon here are:

- Small – less demand, fewer facilities, fewer modes of transport, fewer stakeholders and fewer local impacts; often suburban.
- Medium – moderate demand, more facilities, more modes of transport available, more stakeholders and more potential local impacts; located in more urban settings and can provide access to cities.
- City landmark (large scale) – high demand, high number of high-quality facilities, many different modes of transport available, high number of stakeholders from operators, local government and businesses. High potential for local impacts including new offices and housing, regeneration of the area; normally located in the city centre with national, long-distance transport links and in some cases international links. Integration with the surrounding area is essential given the high local impacts.

Figure 5.12 and Table 5.2 provide a summary of the scaling up of services and facilities at an interchange with an increase in interchange size/scale. The facilities provided at a small interchange provide the basis for a minimum level of service. As the size of the interchange facility increases, this minimum level of service should be built upon, with additional facilities to meet the users' requirements of an interchange of such a scale (Green and Hall 2009).

Table 5.2 goes into further detail, providing a checklist of different types of services that can be used by planners in the design stage of an interchange. The checklist indicates the different level of services that should be provided at the three different sizes of interchange. As the size of the interchange increases, with more demand and more local/city activities in

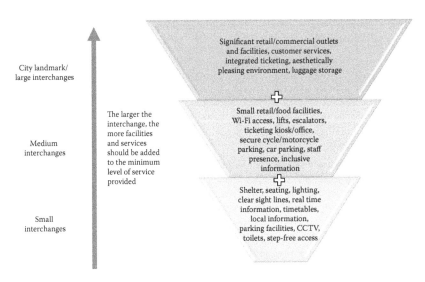

City landmark/
large interchanges

Medium
interchanges

Small
interchanges

The larger the
interchange, the
more facilities
and services
should be added
to the minimum
level of service
provided

Significant retail/commercial outlets
and facilities, customer services,
integrated ticketing, aesthetically
pleasing environment, luggage storage

Small retail/food facilities,
Wi-Fi access, lifts, escalators,
ticketing kiosk/office,
secure cycle/motorcycle
parking, car parking, staff
presence, inclusive
information

Shelter, seating, lighting,
clear sight lines, real time
information, timetables,
local information,
parking facilities, CCTV,
toilets, step-free access

Figure 5.12 Scaling of interchange services and facilities. (Adapted from Green, C. and P. Hall. Better rail stations, Department for Transport. 2009.)

the vicinity, the number of services provided within the interchange should increase also.

Not only should facilities be available to travellers, but they should also be available to others who use the commercial and retail facilities. This will increase revenue for the interchange.

As a result, the interchange would have more or less impacts on the surrounding activities, having an impact on new developments and businesses: shops, jobs, housing and so on. This is a dynamic process where the activities in the local area make the interchange place more important, and subsequently more services and businesses are attracted to locating inside the interchange. This has greater impacts at a local level. The interrelations of the dimensions of the interchange define the degree of integration between activities inside the interchange and those located in the local vicinity. From this, the logical business model and the stakeholders to involve for each particular case can be identified.

Table 5.2 Checklist of recommended facilities to provide at different types of interchanges

Facilities/services	Interchange size		
	Small	Medium	Landmark
Shelter/cover	✓	✓✓	✓✓✓
Seating	✓	✓✓	✓✓✓
Lighting	✓	✓✓	✓✓✓
Ticket machines/kiosk	✓	✓✓	✓✓✓
Real-time information and timetables for the different modes	✓	✓✓	✓✓✓
Local information and maps to support egress from the interchange, particularly by pedestrians and cyclists	✓	✓✓	✓✓✓
Availability of dial-a-ride facilities and information	✓	✓✓	✓✓✓
Information on local taxi services and telephone access	✓	✓✓	✓✓✓
Parking facilities for cars, motorcycles and bicycles with direct, uninterrupted, safe pedestrian routes from car parking	✓	✓✓	✓✓✓
Short transfer distances between modes	✓	✓✓	✓✓✓
Toilets	✓	✓✓	✓✓✓
Help points for customers	✓	✓✓	✓✓✓
Step-free access	✓	✓✓	✓✓✓
CCTV and clear sight lines/good visibility throughout the interchange and surrounding area to provide security	✓	✓✓	✓✓✓
Inclusive information (audible, tactile and for non-native speakers)	✓	✓	✓✓
Staff presence	✓	✓	✓✓
Emergency exits that are clearly indicated	✓	✓	✓✓
Wi-Fi (wireless internet) access		✓	✓✓
Regular public address announcements		✓	✓✓
Retail and food outlets to enable purchase of items for the onward journey		✓	✓✓
Traffic control measures such as speed bumps, pedestrian/cycle crossings in areas of conflict		✓	✓✓
An aesthetically pleasing environment with landscaping and street furniture, where appropriate (should not impede movement) subtle lighting, consistent design and materials			✓
Integrated ticketing facilities and smart readers			✓
Luggage storage			✓
Clear signage between adjacent retail and transport facilities			✓

Table 5.2 (Continued) Checklist of recommended facilities to provide at different
                         types of interchanges

Facilities/services	Interchange size		
	Small	*Medium*	*Landmark*
Designated areas for staff and functions such as deliveries and waste collection			✓
Lifts large enough to carry cyclists and pedestrians			✓
Traffic control measures to prioritise bus movements			✓
Bus movements/facilities that fit with the operation of the surrounding road network			✓
Good legibility for transport users through the organisation of transport modes geographically			✓
Commercial and retail facilities accessible to non-fare-paying users of the interchange			✓

Note: The number of '✓' increases with the requirement for higher-quality facilities.

Chapter 6

Making the interchange attractive for users

Sara Hernandez and Andres Monzon

CONTENTS

6.1 WHAT IS THE USER'S ROLE IN AN URBAN TRANSPORT INTERCHANGE?

As travel patterns become more complex, many public transport users have to make transfers between different modes to complete their daily journeys (Hernandez et al. 2015). It is essential therefore to make interchanges attractive places to transfer in order to achieve or maintain a high level of public transport use. In this respect, measures oriented to improve public transport service quality are required, such as reducing the inconvenience of transfer and providing a seamless travel experience. In this context, the EU Green Paper on urban mobility (European Commission 2007) emphasised that sustainable mobility aims to address three challenges: reducing congestion; improving the quality of public transportation services to achieve a modal shift from private car to public transport; and promoting the soft modes of walking and cycling. Moreover, total travel time directly influences trip choices. Good connectivity at public transport stops and stations is therefore critical to the effectiveness of the overall transportation network (Iseki and Taylor 2010).

Urban transport interchanges play a key role within transport networks since they allow different modes to be used in an integrated manner. Nevertheless, transport stations in general must be considered multimodal facilities where travellers are not only passing through; they are also spending time there (van Hagen 2011). For this reason, public transport users are particularly affected by the quality of the service provided. Terzis and Last (2000) highlighted that transport interchanges should meet sustainability standards and also be attractive for users given that their physical experiences and psychological reactions are significantly influenced by the design and operation of the interchange. In this context, capturing the user's experience and perceptions is crucial to achieve the most appropriate policy measures for public transport interchanges.

6.1.1 Perceived quality of transport services and their facilities

Although several studies have been undertaken to evaluate users' satisfaction with the quality of public transport services (e.g. dell'Olio et al. 2011), evaluation methods have rarely been adapted for measuring users' requirements, as well as their perceptions of the quality of urban transport interchanges during their operation (Dell'Asin et al. 2014; Hernandez et al. 2015).

Consequently, an ad hoc travellers' attitudinal survey was designed and implemented in five European interchanges in order to capture user's views and perceptions related to different aspects and elements of an interchange. The results obtained from the analysis are used to validate findings, improve the experience of intermodal trips and provide inputs for the City-HUB model.

In this respect, this section aims to outline the key factors for users that define an efficient urban transport interchange. Likewise, a novel methodological framework is developed which allows interchange managers to formulate appropriate strategic decisions oriented not only to enhance the transfer experience but also to improve the quality of service. Both of the main findings presented and the methodology proposed are based on two studies developed by Hernandez and Monzon (2016) and Hernandez et al. (2015), respectively.

6.2 TRAVELLERS' ATTITUDINAL SURVEY TO CAPTURE THE USER'S EXPERIENCE

The quality of service provided in a transport interchange has a direct influence on traveller experience and perception and the best way to know and understand their views and needs is through a travellers' attitudinal survey. To this end, an ad hoc survey applicable to any transport interchange was designed in the City-HUB project (see Appendix III).

Since several aspects are relevant to the traveller's decision-making, related to their socioeconomic characteristics and their travel habits, the survey was structured in three different parts (Figure 6.1).

Figure 6.1 Structure of the City-HUB travellers' attitudinal survey.

6.2.1 Part A: travellers' satisfaction questionnaire

This part contained 37 items relating to various aspects and elements of an interchange, classified in eight different groups (see Figure 6.2). These aspects and elements are based on the findings from the literature review (Pirate 2001; Terzis and Last 2000; Durmisevic and Sariyildiz 2001; Iseki and Taylor 2010; Abreu e Silva and Bazrafshan 2013). Moreover, one question about the overall satisfaction with the interchange was included. Each respondent was asked to rate their satisfaction level with each of the items

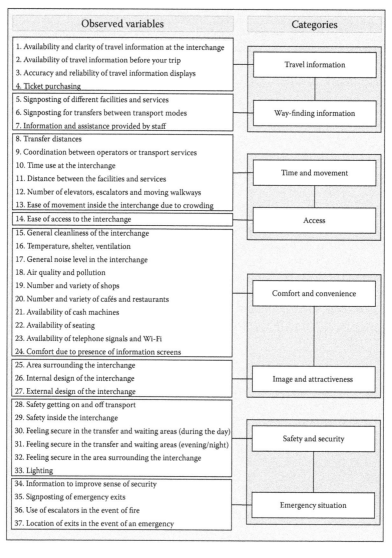

Figure 6.2 Part A: City-HUB travellers' satisfaction questionnaire.

and their overall satisfaction using a 5 point scale (a Likert scale) from 1 (strongly dissatisfied) to 5 (strongly satisfied).

6.2.2 Part B: trip habits

This part gathered some information about users' trip habits. The most relevant information is related to:

- Trip purpose
- Selected transport mode (from the origin to the interchange and from the interchange to the destination)
- Time (to/from/inside the interchange)
- Use of time inside the interchange
- Type of public transport ticket used

6.2.3 Part C: socioeconomic characteristics

This part aimed to define the user profile. Therefore, this part collected some questions related to socioeconomic characteristics, such as:

- Gender
- Age
- Education level
- Employment status
- Household size
- Household net income
- Driving license
- Access to private vehicle (car, motorcycle, bicycle)

6.3 SURVEY AT URBAN INTERCHANGES: DATA COLLECTION METHOD

Traditional methods such as face-to-face interviews are not considered appropriate at urban transport interchanges. They are nodes in the public transport network where users transfer rapidly from one mode of transport to another. Therefore, users would be unlikely to participate in a 20 min questionnaire that would interrupt their journey unnecessarily (Hernandez et al. 2015). The City-HUB project proposes a new implementation method to improve data collection rates and reflect the realities of urban interchanges. This approach combines computer-assisted methods with traditional survey techniques (a face-to-face distribution process). This data collection method is described in detail by Hernandez et al. (2015). The survey was carried out online via SurveyMonkey (a software platform to conduct surveys), and a prize draw was offered for participants.

To capture the user's attention, a card marked with a 'Survey number' was handed out to travellers with the following information: survey objectives and project description; link to the survey website; and information on the prize draw. The 'Survey number' provided each respondent with an individual access code to the online survey which was accessible on computers, smartphones and tablets. Additionally, each person who received this card was recorded in a control sheet in order to subsequently check that the respondent was the traveller who received the card at the interchange. This control sheet contained the following information: 'Survey number', date, time, location in the interchange, group of population (gender and age) and observations (luggage, pushchair, etc.).

Lastly, the answers recorded by the SurveyMonkey tool were directly exported to a database avoiding input errors or illegible responses. This new approach allowed the survey teams to reach more users in less time, therefore increasing the sample sizes.

6.4 DESCRIPTION OF THE PILOT CASE STUDIES

A set of five urban transport interchanges were selected as pilot case studies in order to capture users' views and experience, and to assess good and bad practices, obstacles and potential improvements from the daily operations of existing public transport interchanges. They were selected with consideration given to a balance of geography and heterogeneity in terms of modes, ownership structure and size. The case studies selected were:

- Moncloa Interchange (Madrid, Spain)
- Kamppi Interchange (Helsinki, Finland)
- Ilford Railway Station (London, United Kingdom)
- Köbánya-Kispest Interchange (Budapest, Hungary)
- Railway Station of Thessaloniki (Thessaloniki, Greece)

The location as well as the transport modes involved are key variables to determine the category of a multimodal transport station (Pitsiava-Latinopoulou and Iordanopoulos 2012). Therefore, all of the interchanges selected play a key role in multimodal trips in their corresponding cities (see Table 6.1). They cover a wide spectrum of interchange types and geographical distribution.

6.4.1 Moncloa Interchange (Madrid, Spain)

Moncloa Interchange is located on the north-western edge of Madrid, Spain, providing a gateway to the city for over 287,000 people a day. It mainly hosts local and regional bus services, as well as a few national bus lines. Moreover, Moncloa is the metro station with the highest daily demand in Madrid.

Table 6.1 Main characteristics and business models of the selected pilot case studies

Urban transport interchange	Year built	Role	Modes of transport	Daily passenger demand
Moncloa, Madrid (Spain)	1995, refurbishment in 2008	Local Regional National	56 metropolitan bus lines 20 urban bus lines 2 metro lines 2 long-distance bus lines	100,000 53,000 134,000 Subtotal: 287,000
Kamppi, Helsinki (Finland)	2005 (started operation) 2006 (shopping centre)	Local Regional National International (1 bus to St. Petersburg)	21 local bus lines 40 regional bus lines 15 metropolitan bus lines 1 metro line 1 international bus line 2 tram lines Cycle centre Car parking Taxi	19,500 8,500 7,500 21,700 Subtotal: 57,000
Ilford Railway Station, London (UK)	1839 Rebuilt in 1980 Planned for refurbishment	Local Regional	Rail Bus Cycle (with cycle parking) Drop off and car parking Taxi	21,000 Subtotal: 21,000
Kőbánya-Kispest, Budapest (Hungary)	Refurbishment in 2008 Rail station not refurbished since 1980	Local Regional National International	Rail Metro Local buses Suburban buses Bus line to the airport Cycle Park-and-ride	10,000 68,000 74,500 3,000 Subtotal: 155,500
Railway Station of Thessaloniki, Thessaloniki (Greece)	1961	Local Regional National	Regional rail Urban bus Suburban bus Bus line to the airport Cycle Park-and-ride, kiss-and-ride Taxi Metro (under construction)	6,000 138,000 22,500 Subtotal: 166,500

Figure 6.3 Moncloa Interchange – Bus island 2. (Courtesy of Andres Garcia.)

This interchange has four entrances and is distributed over four levels, and the bus area is divided into three different islands with a total of 39 bus bays (Figure 6.3). Metropolitan buses are distributed in each island according to common routes. Likewise, bus lines are allocated to the same bus bay each time, so that travellers know where they arrive or depart from. Furthermore, each island connects to the lower level, where the metro station entrance hall, travel services (information desk, ticket purchase, etc.) and retail area are located.

Moncloa Interchange is equipped with appropriate signs and screen displays. Moreover, routes to be followed through the station are identified by different colours on the floors, walls and ceilings, making it easier to identify different areas. The public transport authority is responsible for the ticket and fare integration for the public transport regional network in Madrid.

6.4.2 Kamppi Interchange (Helsinki, Finland)

Kamppi Interchange is situated in the basement of a new shopping centre in a central location in downtown Helsinki. The area of the interchange/shopping centre is approximately four hectares. In addition to the shopping centre, there are also offices and flats in the same building (Figure 6.4).

Modes of transport at the interchange include local, regional, national and international buses (to St. Petersburg, Russia), metro, tram, bicycle, car and taxi. The average number of visitors to Kamppi on working days is approximately 100,000, of which 84,000 use public transportation. Furthermore, the central railway station is approximately 500 m away from the Kamppi interchange. There are also bus stations for most of the northern and eastern local and regional buses located adjacent to the railway station.

Figure 6.4 Kamppi Interchange. (Courtesy of Andres Monzon.)

The ticket system is integrated for local and some regional transport in buses, trams, metro and trains. However, long-distance buses use different fares. Likewise, timetable information is available on general displays as well as on individual displays at each gate – different transport operators share the same information screens.

6.4.3 Ilford Railway Station (London, United Kingdom)

Ilford is a large suburban town in the London Borough of Redbridge (East London). The Ilford Railway Station is situated on the Great Eastern Main Line and has regular local train services (from Essex) to Liverpool Street Station in central London. More than ten bus stops are located within walking distance of the station, with the town being a hub for the London Buses network.

This station is predominantly a railway station, with bus services on the street. The modes of transport available at the interchange are: main-line rail, bus, cycle (with cycle parking), pedestrian, private car with drop-off, car parking and taxi. Additionally, there is a shopping centre opposite the station. The station is used mainly for local commuter trips, with the main access modes being buses and walking. The station has two entrances. At the main entrance there is a small ticket hall, with a gate through which passengers pass to access the five platforms. The majority of passenger facilities are found on Platforms 2 and 3, including two indoor waiting areas, seating, a customer information office, a newsagent, vending machines and toilets.

A particular issue at Ilford is that currently there is no way of accessing the platforms other than by using the stairs, as a result of the stair-lifts being broken and unable to be fixed due to the parts being unavailable. In

Figure 6.5 Ilford Railway Station. (Courtesy of Jan Spousta.)

this respect, the interchange is planned for redevelopment as part of the Crossrail project. The station improvements will provide a new ticket hall layout with greater gate line capacity, passenger lifts, longer platforms and a realigned station entrance and elevation to the street (Figure 6.5).

6.4.4 Köbánya-Kispest (Budapest, Hungary)

Köbánya-Kispest Interchange is one of the backbones of public transport in Budapest connecting the northeast and southeast of the city via the city centre. This interchange includes a metro terminal, bus terminal, railway terminal, park-and-ride and a shopping mall. The transport modes involved are a metro line (M3), 15 local bus lines, 3 regional bus lines and 2 railway lines. Although all types of connections are available, the interchange primarily handles local and suburban traffic. Moreover, Köbánya-Kispest is a major railway station (2 main-line railways).

The interchange was fully refurbished in 2011, with the exception of the railway station (which for rail passengers has meant a lack of attractiveness and poor accessibility for disabled people). All transfer routes among terminals are covered and protected from the weather. The stations and platforms are open air, but covered, so passengers are protected from the rain, but not from the cold. There is an overpass which is noise protected and is used as a waiting area (Figure 6.6).

Travel information is not uniform because of the existence of different managers and operators. Furthermore, there is no integrated ticketing – separate tickets have to be bought for the services of each transport operator.

Figure 6.6 Köbánya-Kispest Interchange. (Courtesy of Jan Spousta.)

6.4.5 Railway Station of Thessaloniki (Greece)

The so-called New Railway Station is the central passenger interchange terminal in Thessaloniki, situated close to the city business district allowing for the movement of travellers around the city. The station also works as a terminal for public bus services, and it is directly connected to the interurban bus station, where scheduled departures for Athens and other Greek cities are available (Figure 6.7). Moreover, the station is close to the port of Thessaloniki, while a bus line connects the station with the international airport of Thessaloniki, 'Macedonia'. Taxis, bicycle ways, park-and-ride, kiss-and-ride and metro (under construction) are modes also available at the station. The construction of the metro station will especially reform both the terminal and the surrounding area, transforming the current terminal into a modern integrated bus–railway–metro station in the future.

The station lacks a cooperative management structure, since each of the involved transport operators is responsible for the management of the space and the operations assigned to it. Furthermore, an integrated ticketing policy is lacking. Nevertheless, it has to be noted that integrated cooperation among the different operators is present in terms of timetables (scheduling).

Figure 6.7 Railway Station of Thessaloniki. (Courtesy of Jan Spousta.)

Also, on a pilot basis, an effort is being made for the combination of trains and taxis within the framework of integrated door-to-door movement with the same ticket.

6.5 SURVEY IMPLEMENTATION PROCESS: RESULTS AND DATA COLLECTED

The surveys were carried out between May 2013 and October 2013. They started and finished at different times, and at each case study, the online survey was subsequently open on the web for around one month for receiving travellers' answers. In the case of Thessaloniki, it was necessary to repeat the process because of the low response rate. On the other hand, the survey carried out at Ilford Railway Station combined this new implementation method with a traditional postal survey (Figure 6.8).

The highest response rates (see Table 6.2) were reached in Moncloa and Kamppi, with values of 23.2% and 16.5%, respectively. The lowest

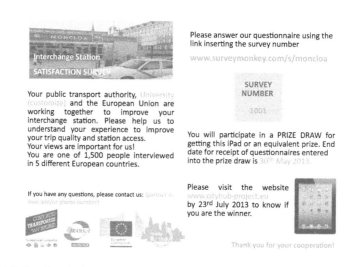

Figure 6.8 Card handed out at Moncloa Interchange, Madrid, Spain.

Table 6.2 Response rate in each pilot case study

Case study	Survey tool	Cards delivered	Received surveys	Valid surveys	Response rate
Moncloa	Online	4,000	928	865	23.2%
Kamppi	Online	1,874	309	298	16.5%
Ilford Railway Station	Postal/online	1,600	165/69	226	14.1%
Köbánya-Kispest	Online	4,100	383	383	9.4%
Railway Station of Thessaloniki	Online (2 periods)	2,000	105/153	244	12.9%

response rate was for Köbánya-Kispest, where a large number of cards were also delivered (4100) – 9.4% with 383 surveys received. A minimum sample of 200 surveys was reached for each case study, with 226 surveys received in Ilford, up to 928 in Moncloa. The prize for each location was an Apple iPad (iPad 2 or iPad mini), except in Thessaloniki, where ten monthly transport cards were provided.

6.6 CROSS-SITE COMPARISON: USER PROFILES AND TRIP PATTERNS

To define and analyse the key factors in the design, operation and management of an interchange, or to analyse their potential strengths and weaknesses, it is first necessary to understand the user profile and trip habits in each pilot case study. This section therefore presents results obtained from Part C and Part B of the survey, respectively, such as users' socioeconomic information, including gender, age, employment status and household net income, and information related to users' travel habits, including frequency of use of the interchange, the purpose of the trip, the selected transport mode (from origin to the interchange and from the interchange to destination) and times (to/from/inside the interchange) at each pilot case study.

6.6.1 Part C: what is the user profile?

Figure 6.9 shows the characteristics of each sample by gender and age. It was found – see Figure 6.9a – that 64.1% and 59.8% of the respondents were female in Kamppi and Thessaloniki, respectively, whereas in Ilford only 39.9% of the respondents were female. Moncloa and Köbánya-Kispest had a more homogenous sample of about 50%. There was no representative sample of younger users (those under 18 years old) or older users (those above 65 years old) in any case study. Therefore, users from all case studies were between 18 and 65 years old. At Ilford and Kamppi interchanges, most users were between 26 and 65 years old.

Regarding the level of household net income per month, and in order to compare all results of the five case studies, three different levels were considered, taking into account the minimum revenue or alternative measure of revenue of each country. Therefore, the minimum level of household net income per month is less than or equal to two times the minimum revenue; the average level is between two and four times the minimum revenue; and the maximum level of household net income per month is more than or equal to four times the minimum revenue. Figure 6.10 shows that Kamppi has a higher proportion of users with the highest level of household net income per month (34.5% of the sample), whereas Thessaloniki has a higher proportion of users with the lowest level of household net income per month (56.4%). The remaining case studies have a more equal distribution.

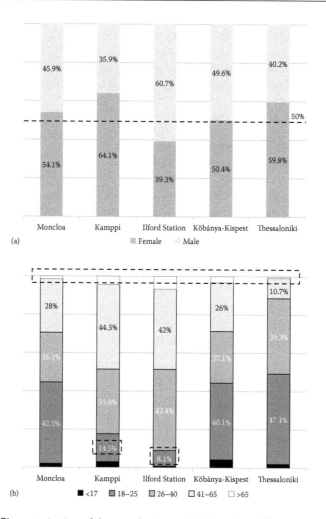

Figure 6.9 Characterisation of the samples by gender (a) and age (b).

By and large, the travellers surveyed are habitual users (i.e. they use the interchange daily or 3–4 times a week), as shown in Figure 6.11. Moncloa has the highest number of habitual users (about 85%). The railway station in Thessaloniki, however, has a different user profile, with most travellers using the interchange few times a month (24%) or less frequently (42%). Kamppi also has a more representative sample of this type of traveller (an infrequent user) with almost 25% using the interchange a few times a month or less. Regarding the trip purpose (see Figure 6.11), work and studies were generally the main purposes for all of the trips recorded, except in the Railway Station of Thessaloniki where the main trip purpose was for leisure (55.7%). Ilford Station is mainly for commuter trips (for local users)

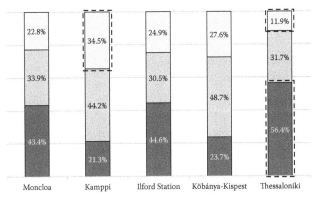

Figure 6.10 Household net income per month.

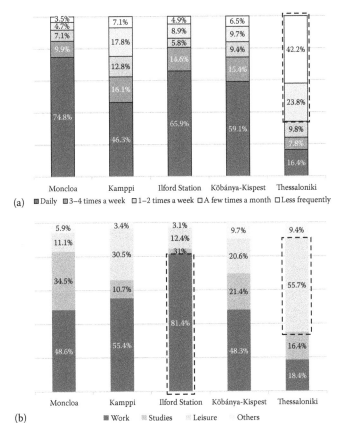

Figure 6.11 (a) Frequency of use of each interchange and (b) trip purpose.

as mentioned earlier and this explains the fact that more than 80% of the respondents were travelling for work.

6.6.2 Part B: what are the trip habits at the interchanges?

In order to understand better the mobility patterns and therefore to identify the role of each urban interchange within the transport network, it is relevant to know the main access and exit transport modes at each interchange case study according to the valid sample of respondents (see Figure 6.12).

The main access mode at Moncloa Interchange is the metro along with the metropolitan buses, which are also the main exit transport mode. Kamppi Interchange is situated in a very central location in Helsinki and, as mentioned earlier, it connects various transport modes (buses, metro, tram, bicycle, car and taxis). The main access and exit modes at Kamppi, from the sample of users interviewed, are long-distance and metropolitan buses, perhaps because the surveys were handed out in three different terminals and the number was not the same in each of them. The walking mode is also important in both cases (both access and exit). Ilford Station is predominantly a railway station (main exit transport mode) and the main access modes are buses and walking. In the case of Köbánya-Kispest, the metro plays a key role in the interchange as both the main access and exit mode. Finally, the Railway Station of Thessaloniki involves several transport modes apart from rail. The urban bus is the main transport mode for access and exit to the station.

Finally, the analysis looks at the users' travel time. The results are expressed as the median (minutes). Table 6.3 shows these results according

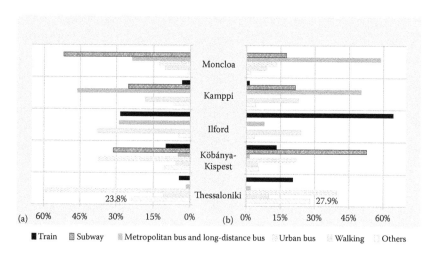

Figure 6.12 Main access (a) and egress (b) transport modes at each pilot case study.

Table 6.3 Characterisation of user travel time at each pilot case study expressed with the median (minutes)

Case study	Time TO interchange[a]	Time INSIDE interchange	Time FROM interchange[b]	Total travel time
Moncloa	20	10	30	50
Kamppi	20	9	20	45
Ilford Railway Station	15	6	30	45
Köbánya-Kispest	20	8	25	50
Railway Station of Thessaloniki	25	–	30	120

[a] From the origin to the interchange.
[b] From the interchange to the final destination.

to the time to/from the interchange, time inside the interchange and total travel time. The Railway Station of Thessaloniki has the highest average travel time, approximately 2 h; however, there was no information provided about the time spent inside the interchange and how users spend this time.

To make an interchange attractive for users, it is crucial to find out not only how long they spent inside the interchange, but also how many users are spending time there, and how they use this time in order to allocate and manage resources better. Time spent at interchanges could be distributed between different activities such as queuing, transferring, shopping and other pursuits (Figure 6.13).

The time spent for each activity at each pilot case study is shown in Table 6.4 (except in the Railway Station of Thessaloniki due to lack of data). The shortest queuing times were observed at Kamppi and Köbánya-Kispest interchanges, whereas Moncloa transport interchange records the highest queuing times (41% of users wait for between 5 and 15 min and

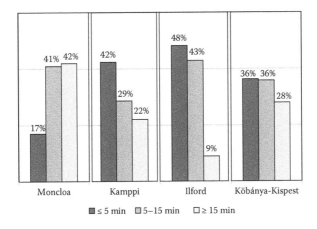

Figure 6.13 Time spent inside the interchange for each pilot case study.

Table 6.4 Percentage of users according to their waiting time by time intervals and activities

Case study	Queuing			Transferring[a]			Shopping			Other activities[b]		
	≤5 min	5–15 min	≥15 min	≤5 min	5–15 min	≥15 min	≤5 min	5–15 min	≥15 min	≤5 min	5–15 min	≥15 min
Moncloa	39.0	**41.0**	**20.0**	**71.0**	**13.0**	**16.0**	**92.0**	6.0	2.0	**82.0**	13.0	6.0
Kamppi	**86.1**	9.4	4.5	67.3	12.9	19.8	53.6	27.0	**19.4**	71.4	9.2	**19.4**
Ilford	73.2	21.1	5.6	86.7	2.2	11.1	68.3	**29.0**	2.8	75.6	**15.4**	9.0
Kőbánya-Kispest	**86.3**	13.7	–	72.0	9.8	18.2	77.6	17.8	4.7	**85.8**	6.3	7.9

[a] Walking between different transport modes for boarding.
[b] Buying transport tickets, seated at café or seated at waiting areas, etc.

20% more than 15 min). On the other hand, the shortest transfer time is in Moncloa, being less than 5 min. Conversely, Kamppi Interchange has the highest transfer time (19.4% of users have a transfer time longer than 15 min). Regarding other activities (e.g. buying transport tickets, seated at a café or seated in waiting areas, etc.), users from Ilford Station spent less time than users from Kamppi and Köbánya-Kispest. Finally, users who spend more time shopping are from Kamppi Interchange compared with users from Moncloa and Köbánya-Kispest interchanges who spend less than 5 min.

6.7 PART A: EXPLORING THE USER EXPERIENCE AND PERCEPTIONS: ASSESSMENT PROCESS

The data collection process is crucial to properly set the context of an urban transport interchange as well as for the analysis of socioeconomic information and trip habits collected in the survey. Figure 6.14 shows this procedure which has been followed in the previous sections. Once the user profiles, along with their trip habits, have been identified, it is possible to better understand their perception of the interchange and their reasons for this view. These previous steps will be inputs for defining and developing the 'Assessment process'.

This section aims to provide an assessment process to managers and developers of urban transport interchanges for making them more attractive for users. To this end, a methodological framework is proposed, as shown in Figure 6.15. The 'Assessment process' generates answers to key questions oriented towards the management and planning of an urban transport interchange, respectively: How is an existing transport interchange

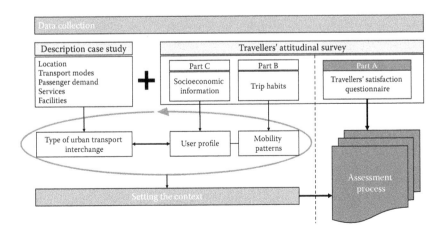

Figure 6.14 Steps to set the context of an urban transport interchange.

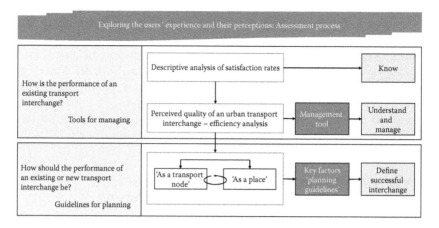

Figure 6.15 Assessment process for exploring the users' experience and their perceptions.

performing? How should an existing or new transport interchange be performing? Additionally, the assessment process, based on the user's experience and perceptions, provides managers and developers with a depth of knowledge of the interchange in order to better understand its performance and manage the available resources properly. Lastly, the key factors of an urban interchange identified under a dual approach – 'as a transport node' and 'as a place' – define a successful and attractive interchange.

6.7.1 Descriptive analysis of satisfaction rates

A descriptive analysis of Part A of the survey – 'Traveller satisfaction questionnaire' – allows us to know the user's perceptions of the performance of an existing transport interchange. This first step of the assessment process reveals the user's satisfaction level with regard to different aspects and elements of an interchange related to transport services and facilities.

In this respect, the results obtained in the pilot case studies reveal that users are generally satisfied, with an average overall satisfaction higher than 3 in a 5 point scale, with high average satisfaction rates in Kamppi (3.94), Moncloa (3.92) and Köbánya-Kispest (3.61) interchanges, and acceptable average rates in Ilford Railway Station (3.16) and the Railway Station of Thessaloniki (3.13). Regarding the satisfaction rates by gender and age groups, no significant differences were found. Table 6.5 shows the aspects of the interchange most and least valued by users. 'Access' to interchanges achieves the highest average rate in all cases, except at Ilford Station, where 'Travel Information' occupies the top position. At Moncloa, Kamppi and Köbánya-Kispest interchanges, all dimensions rate higher than 3.00, so it could be said that travellers are satisfied with each aspect of these

Table 6.5 Satisfaction rates aggregated by categories

Categories	Moncloa	Kamppi	Ilford	Köbánya-Kispest	Thessaloniki
Travel Information	3.69	3.92	3.44	3.61	3.37
Way-finding information	3.81	3.50	3.26	3.70	3.26
Time and movement	3.69	3.69	3.09	3.52	3.44
Access	4.19	4.29	3.33	4.32	3.73
Comfort and convenience	3.35	3.59	2.99	3.41	2.95
Image and attractiveness	3.77	3.27	2.50	3.56	2.16
Safety and security	3.75	3.84	3.10	3.59	2.72
Emergency	3.45	3.42	2.88	3.70	2.90

interchanges. On the other hand, users from Ilford and Thessaloniki are not very satisfied with some of these aspects, such as 'Comfort and Convenience', 'Image and Attractiveness', 'Emergency situations', as well as, in the case of Thessaloniki, 'Safety and Security'. These categories scored less than 3.00.

Moreover, Figure 6.16 shows users' overall satisfaction for each pilot case study, where scores of 1 and 2 are categorised as 'Dissatisfied', a score of 3 as 'Normal' and scores of 4 and 5 as 'Satisfied'. At Moncloa and Kamppi interchanges, the majority of users feel satisfied. Conversely, at the railway stations of Ilford and Thessaloniki, there is a large percentage of users feeling 'Dissatisfied' with the interchanges (more than 20% in both cases). At Köbánya-Kispest, the percentage of 'Satisfied' users diminishes, but they feel a 'Normal' satisfaction level.

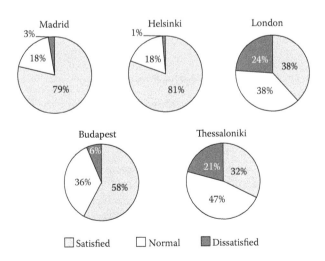

Figure 6.16 Overall satisfaction case by case.

6.7.2 Tool for assessing the perceived quality in an urban transport interchange

Some studies have begun to investigate the quality of public transport infrastructures due to a growing interest in developing smart and efficient facilities for intermodal transfers. Guo and Wilson (2011) concluded that improving transfer experience could significantly benefit public transport. However, evaluation methods have rarely been adapted for measuring user needs or their perceptions of the quality of urban transport interchanges (Dell'Asin et al. 2014; Hernandez et al. 2015).

This section goes a step further and briefly presents a novel and useful tool, developed by Hernandez et al. (2015), for assessing the perceived quality in an urban transport interchange. This tool not only analyses the user's satisfaction concerning different aspects and elements of an urban transport interchange, but it also identifies the 'derived importance' of each of them, thus identifying the potential strengths and weaknesses of an urban transport interchange.

This technique, based on a two-step analytical procedure, combines two powerful methodologies:

- Classification and Regression Trees model (CART model):
  Objective: obtaining the 'derived importance' of each aspect and element of an interchange
- Importance-Performance Analysis (IPA):
  Objective: identifying which aspects or elements of an interchange need improvement and which can be considered strengths

This methodological framework allows interchange managers to formulate appropriate strategic decisions oriented not only to enhance the transfer experience but also to improve the quality of service at urban transport interchanges (Hernandez et al. 2015).

6.7.2.1 Step 1: obtaining the 'derived importance'

The first step of this analytical procedure aims to deduce the importance – 'derived importance' – of each aspect and element collected in the survey which are considered as independent variables (see Figure 6.2). The technique applied to achieve this purpose is the CART model, developed by Breiman et al. (1984), which reflects the impact of the predictor variables on the model (Kashani and Mohaymany 2011).

As defined by Hernandez et al. (2015), the basic CART concept is to divide a dataset into 'purer' subsets.* The process begins with a dataset

---

* A subset is considered pure when it contains observations belonging to a single class (Breiman et al. 1984).

concentrated in a root node. Then, a set of split rules is established where all independent variables included in the analysis are possible splitters.* The splitting process is applied recursively to each child node until all data included in the node are of the same class (i.e. the node should be considered 'pure'), their homogeneity cannot be improved or a stopping criterion has been satisfied. Finally, in order to reduce the resulting tree complexity, a cost–complexity pruning algorithm is applied to remove branches that add little to the predictive value of the tree. The final step is therefore to select an 'optimal tree' from the pruned trees which determines the derived importance of the variables. For that, the original dataset is usually divided into two subsets, one for 'learning' or 'growing the tree' and the other for testing or 'validation'. Consequently, the optimal tree offers the lowest misclassification cost for test data (de Oña et al. 2012).

Figure 6.17 shows the optimal tree obtained at Ilford Railway Station along with the interpretation of the analysis. The root node (Node 0) is divided into two child nodes (Node 1 and Node 2). The splitter (independent variable that maximises the 'purity' of the two child nodes) is: 'Safety (inside interchange)'. The majority of the sample (62.4%) is concentrated in the child Node 1, which indicates that the splitter is a major discriminant of the model. Moreover, when 'Safety (inside interchange)' is rated with a score equal to or higher than 4 (see Node 2), the users have a probability of 68.2% of feeling satisfied. The splitter criterion is therefore applied recursively until the optimal tree is obtained. During the process, other independent variables act as splitters of the tree, such as 'Travel information at interchange'. After obtaining the optimal tree, the 'derived importance' is obtained for each of the independent variables included in the CART model. The variable used as the dependent variable is 'Overall satisfaction' (question from the Part A of the survey).

6.7.2.2 *Step 2: importance-performance analysis*

This second step is oriented to managers and developers of transport interchanges. Importance-Performance Analysis (IPA) is a graphic technique that uses the importance and performance ratings of different aspects of a service as coordinates on a two-dimensional grid divided into four quadrants (Martilla and James 1977). This quadrant chart quantifies the 'derived importance' assigned to each attribute (vertical axis) and shows the users' ratings of the quality of each element (horizontal axis). The horizontal and vertical axis are determined as average values of derived importance, previously obtained in Step 1, and satisfaction ratings, respectively. Additionally, a performance threshold is established and, therefore, attributes with satisfaction ratings below this threshold should be

---

* An independent variable is considered as a splitter when it is able to create the greatest homogeneity in the child nodes (Hernandez et al. 2015).

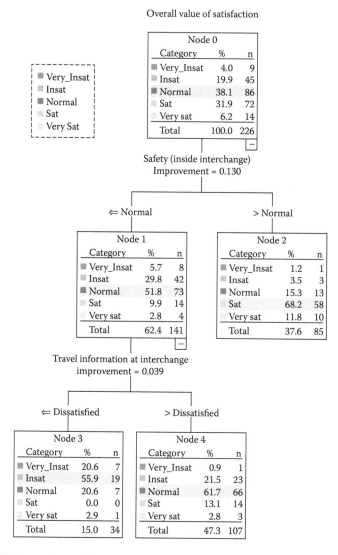

Figure 6.17 Optimal tree of Ilford Railway Station.

improved immediately without further evaluation. The position of each attribute within the four quadrants of the IPA matrix shows the relative urgency of improvement.

The outcomes obtained at Ilford Railway Station (see Figure 6.18) show that most of the attributes were perceived as being of low quality – below the performance threshold. In this case, such a performance threshold coincided with the average value of satisfaction rates (*y*-axis).

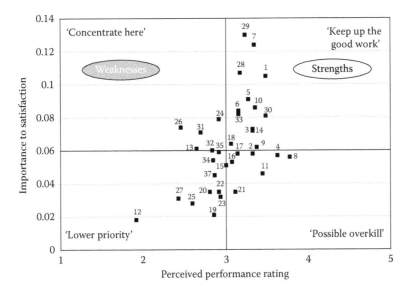

Figure 6.18 Importance-performance analysis of Ilford Railway Station according to users' perceptions.

- *'Concentrate here' quadrant:* This quadrant contains the characteristics which users feel are important but with which they are dissatisfied. It implies that improvement efforts should be concentrated on these aspects of the interchange. Aspects such as 'Comfort due to the presence of information screens', 'Internal design of the interchange' and 'Feeling secure in transfer/waiting areas in the evening/night' should urgently improve. They are the potential weaknesses of the interchange. Likewise, 'Number of elevators, escalators and moving walkways' is the aspect worst valued by the users.
- *'Possible overkill' quadrant:* This area indicates attributes which do not have a strong influence on users' evaluations but nevertheless are perceived to be of high quality; transport managers should consider allocating their resources to improvements elsewhere. 'Ticket purchase' and 'Transfer distance between different transport modes' represent overkill in the interchange.
- *'Keep up the good work' quadrant:* This section groups attributes which are considered to be of high importance to users and which are also ranked high in terms of performance, indicating that they are the greatest strengths of the interchange. At Ilford Railway Station, elements related to travel information, time and movement, way-finding information and safety and security are considered the potential strengths of the interchange.

- *'Lower priority' quadrant:* Other attributes that transport managers should work on to increase users' satisfaction are allocated to this quadrant. These elements do not play a key role in users' overall evaluation, but their level of quality is quite low. At Ilford Railway Station, these aspects are 'Ease of movement inside the interchange', 'Number and variety of shops and cafés', 'Surrounding area' and 'External design' of the interchange.

6.7.2.3 What are the users' perceptions of efficiency in each pilot case study?

This tool enables us to evaluate the performance level of an interchange and its efficiency. The efficiency of an interchange should be analysed as a whole, that is, considering not only the users' satisfaction but also understanding how important the aspects and elements of an interchange are. This way, it is possible to measure the performance of an interchange and efficiently manage the available resources. Following the application of this methodological framework, some similarities can be extracted from among the pilot case studies (see Table 6.6). Regarding the average value of the attributes' performance rating (y-axis), Moncloa, Kamppi and Köbánya-Kispest obtained an average value higher than 3.5, while Ilford and Thessaloniki reached an average rating of around 3 (performance threshold established below which the level of quality is unacceptable).

6.7.2.4 Potential weaknesses

The aspects and elements valued below the performance threshold at Moncloa and Kamppi are 'Availability of cash machines and seats', and additionally, 'Availability of telephone signal and Wi-Fi' in the case of Moncloa. All of these factors are related to the 'Comfort and Convenience' category (see Figure 6.2). Additionally, their potential weaknesses are mainly related to aspects of 'Emergency situations' in the case of Moncloa, and 'Image and Attractiveness' at Kamppi interchange. On the other hand, the potential weaknesses at Köbánya-Kispest interchange differ from the previous ones. They are more related to 'Air quality', 'Feeling secure in transfer/waiting areas' and other aspects, such as 'Cleanliness' and 'Design' of the train station. On the contrary, at Ilford and Thessaloniki railway stations, a large number of attributes are perceived with a satisfaction level lower than 3. Their potential weaknesses are related to 'Image and attractiveness' as well as 'Safety and Security'. Furthermore, elements related to 'Comfort and convenience' at both interchanges do not reach the performance threshold established.

Table 6.6 Efficiency analysis of each pilot case study

	Moncloa	Kamppi	Ilford Railway Station[a]	Köbánya-Kispest[b]	Railway Station of Thessaloniki
Aspects and elements affecting user experience					
1. Travel information at interchange			S		S
2. Travel information before trip					S
3. Accuracy of travel information displays					
4. Ticket purchasing				W	
5. Signposting facilities and services	S				
6. Signposting for transfers	S				
7. Information provided by staff			S		
8. Transfer distances					
9. Coordination between different operators					
10. Time use at the interchange					S
11. Distance between facilities and services					
12. Number of elevators, escalators					
13. Movement inside the interchange		W	W		
14. Ease of access to the interchange					
15. General cleanliness of the interchange				W	
16. Temperature		S			
17. General noise level in the interchange					
18. Air quality and pollution					
19. Number of shops					
20. Number of cafés and restaurants					
21. Availability of cash machines					
22. Availability of seating					
23. Telephone signals and Wi-Fi					
24. Comfort due to information screens	W		W	W	W
25. The area surrounding the interchange		W			W
26. The internal design of the interchange		W	W	S	

(Continued)

Table 6.6 (Continued) Efficiency analysis of each pilot case study

	Moncloa	Kamppi	Ilford Railway Station[a]	Köbánya-Kispest[b]	Railway Station of thessaloniki
27. The external design of the interchange		W		W	
28. Safety getting on and off transport		S	S		
29. Safety inside the interchange	S	S	S	S	
30. Feeling secure in transfer and waiting areas (during the day)	S	S		S	
31. Feeling secure in transfer and waiting areas (evening/night)			W		
32. Feeling secure in the area surrounding					
33. Lighting				S	W
34. Information to improve your sense of security	W				W
35. Signposting of emergency exits					S
36. Use of escalators in the event of fire	W				
37. Location of emergency exits	W				
Efficiency analysis					
Aspects and elements below the satisfaction threshold (less than 3)	3	2	15	10	18
Aspects and elements above the average satisfaction rate	19	21	20	28	19
Aspects and elements below the average satisfaction rate	18	16	16	17	18
Satisfaction Ratio	**1.05**	**1.31**	**1.25**	**1.64**	**1.05**
Aspects and elements above the average importance rate	20	18	18	22	19
Aspects and elements below the average importance rate	17	19	18	23	18
Importance ratio	**1.18**	**0.95**	**1.00**	**0.96**	**1.05**
Interchange performance	☺	☺	☹	☹	☹

[a] Element 36 was not included in Ilford Railway Station because it does not have escalators.
[b] The Köbánya-Kispest survey included 45 aspects (some of them were asked for each type of terminal).
Ilford railway station, Köbánya-Kispest and the railway station of Thessaloniki are not considered successful interchanges because there are so many aspects and elements below the satisfaction threshold (less than 3). Therefore, overall perceived quality of the interchange is not good independent of the importance rates as well as performance.

6.7.2.5 *Potential strengths*

Elements related to 'Signposting' to different facilities and transport services, 'Internal design', 'Surrounding area' and 'Safety and Security' are considered the potential strengths at Moncloa Interchange. Likewise, 'Access', 'Comfort and Convenience' and 'Safety and Security' are the potential strengths at Kamppi Interchange. The strengths of Köbánya-Kispest Interchange are mainly related to 'Image and Attractiveness' and 'Safety and Security'. Lastly, 'Travel information', 'Use of time inside the interchange' and 'Signposting to emergency situations' are the potential strengths at the Railway Station of Thessaloniki.

6.7.3 Key factors for defining an efficient urban transport interchange

Finally, the literature review and previous experience highlight that there are two types of aspects to take into consideration in the design, operation and management of multimodal passenger transport terminals: 'functional' and 'psychological' features. The ambivalent nature of a transport station creates opportunities for synergies between both connotations: moving and staying (Peek and van Hagen 2002). This section therefore presents the key factors to achieve a successful urban transport interchange as identified under this dual approach. These main findings are extracted from a study developed by Hernandez and Monzon (2016).

Principal Component Analysis (PCA) was applied to the surveys as an analysis methodology to identify the key factors of an urban transport interchange. Through this statistical analysis it is possible to identify latent factors that cannot be directly measured (Field 2009).

The quality perceived by users of a transport interchange, both the infrastructure (building, facilities and so on) and services (information provided, signposting and so on), usually depends on the context of the interchange (Harmer et al. 2014). However, in the analysis of all cases, this research identifies some elements as fundamental. Even though the pilot case studies are significantly different, the key identified factors are common, as well as the main variables observed; that is, the variables with the highest factor loadings.* These key identified factors defining an urban transport interchange are:

- Information
- Transfer conditions
- Safety and security
- Emergency situations
- Design and image
- Environmental quality

---

* Large factor loadings indicate a significant influence of the variable on the latent factor.

- Services and facilities
- Comfort of waiting time

Therefore, they should be considered as fundamental both in the planning process and operation management phases of an interchange. It is worth stressing that 'Environmental quality' is influenced by the context of the urban interchange. It is clustered with the 'Safety and security' factor in the case of the Kamppi interchange, with 'Design' in the case of Köbánya-Kispest and with 'Services and facilities' in the case of the Railway Station of Thessaloniki (Figure 6.19).

According to the results obtained from the analysis, these factors effectively confirm that an urban transport interchange should be defined under two complementary perspectives: 'as a transport node' and 'as a place'.

Aspects related to 'Information' provision – travel information and signposting – and 'Transfer conditions' – distances and coordination between operators – facilitate the use of the interchange as 'a transport node'. Nevertheless, 'Design and Image', aspects related to 'Environmental quality', such as noise, temperature, air quality and the number and variety of shops and cafés are more focused on improving the user's experience at an interchange as 'a place'.

Additionally, availability of telephone signal, Wi-Fi or travel information displays can be oriented to improve the quality of waiting time. Lastly, 'Safety and Security', both inside the interchange and in the surrounding area, are crucial in both approaches and are directly linked to the overall performance of the interchange.

6.8 HOW TO MAKE AN URBAN TRANSPORT INTERCHANGE ATTRACTIVE FOR USERS

Urban transport interchanges are an everyday experience for public transport users and have become, in this respect, fundamental elements in reducing the inconvenience of transfer and improving users' perceptions. Likewise, they should act as a hub in urban areas, designed in a way that attracts people to the facility from the outside area in order to connect them to the transport modes available there (Figure 6.20). However, there are no standards or regulations specifying the facilities and services required in urban interchanges (Hernandez and Monzon 2016) and evaluation methods have rarely been adapted for measuring users' needs and views with regard to multimodal terminals (Dell'Asin et al. 2014; Hernandez et al. 2015).

6.8.1 Data collection

Since the quality of service provided in a transport interchange has a direct influence on the traveller's experience, the best way to know and

Key factor	Moncloa	Kamppi	Ilford railway station	Köbánya-Kispest	Railway station of Thessaloniki
Information	Signposting Travel information (at interchange)	Travel information (displays and at interchange)	Travel information Co-ordination transport operators	Comfort (information screens) Travel information (at interchange) Signposting	Signposting Information provided by staff Travel information (at interchange)
Transfer conditions	Transfer distance Coordination transport operators Use of time (at interchange)	Distance (facilities and transfer)		Coordination transport operators Distance (facilities and transfer)	Distance (between facilities)
Safety and security	Security in transfer and waiting areas Security surrounding area	Safety (inside interchange) Security in transfer and waiting areas Security surrounding area	Security in transfer and waiting areas (evening/night) Security surrounding area	Security in transfer and waiting areas (day) Security surrounding area Safety getting on/off transport mode	Security in transfer and waiting areas Security surrounding area
Emergency situations	Emergency exits (Location and signposting)	Escalators (case of fire) Emergency exits (Location)	Emergency exits (Location and signposting)	Emergency exits (Location and signposting) Information	Emergency exits (Location and signposting)
Design and image	Design (External and internal)	Design (External and internal)	Design (External and internal)	Design (External and internal)	Design (External and internal)
Environmental quality	Air quality Noise	Air quality Temperature	Noise Temperature	Temperature Noise	Noise Comfort (information screens)
Services and facilities	Shops Cafés and restaurants	Cafés and restaurants Shops	Shops Cafés and restaurants	Cafés and restaurants Cash machines	Air quality
Comfort of waiting time[a]	Telephone signal and Wi-Fi Travel information displays	—	—	—	—

[a] Factor 'Comfort of waiting time' has been only identified at Moncloa Interchange (Madrid, Spain).

Figure 6.19 Key factors and their main variables observed case by case.

Figure 6.20 Urban transport interchange as a place. (Illustration by Divij Jhamb.)

understand their views and needs is through a travellers' attitudinal survey. This chapter presents an ad hoc survey along with a new implementation procedure to carry out surveys of interchange users. This new procedure resolves the time problem – users transfer rapidly from one mode to another and the method is unlikely to interrupt their journeys unnecessarily for more than 5 min. On the other hand, it allows researchers to reach more users in less time, therefore increasing the sample sizes. Both the survey and the description of the case study set the context of an urban transport interchange and enable us to better understand users' perceptions and the reasons for their views.

6.8.2 Assessment process

This chapter provides a useful methodological framework for managers and developers for making an urban transport interchange more attractive for users as well as for achieving a successful interchange from the standpoint of management and operations. This 'Assessment process' allows us to answer key questions such as How is an existing transport interchange performing? How should an existing or new transport interchange be performing?

The results obtained from the application of the 'Assessment process' in the pilot case studies led to the following conclusions:

- Some elements generally have a high importance for users regardless of the context of the interchange. These elements are: 'Internal design' and 'Safety and Security inside the interchange'.

- Some characteristics exert little influence on users' overall evaluation, such as 'Transfer distances between different transport modes', 'Distance between facilities and services', 'Number of elevators, escalators and moving walkways', 'Availability of cash machines' and 'Telephone signal and Wi-Fi'. Nevertheless, users' satisfaction with regard to distances, between both transfer and facilities, is generally high. Regarding the availability of cash machines, telephone signal and Wi-Fi, these can be considered top quality factors. If they are not presented, there will not be any negative influence on the users' overall evaluation; however, if they are presented, they could act as aspects that encourage a strong improvement or worsening of users' overall evaluation.
- Although the features and context of each interchange are completely different, the results obtained show key common factors and attributes in all cases. These key common factors define an urban transport interchange under a dual approach: 'as a transport node' and 'as a place'.
- Factors that better define an interchange 'as a transport node' are aspects related to information provision – travel information and signposting – and transfer conditions – distances and coordination between operators. In contrast, design and image, indoor environmental quality, services and facilities, and elements addressed to improve the comfort of waiting time are directly linked the quality of the interchange 'as a place'. Finally, safety and security is of vital importance for users in both approaches (Figure 6.21).

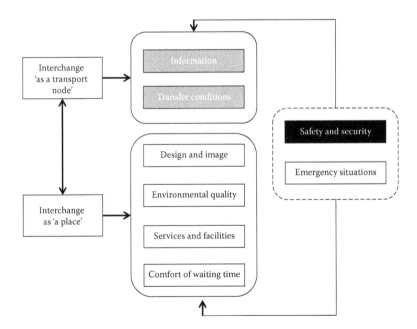

Figure 6.21 Planning guidelines for efficient urban transport interchanges.

Chapter 7

The City-HUB model

Andres Monzon, Floridea Di Ciommo,
Sara Hernandez, Jardar Andersen,
Petter Christiansen and Ricardo Poppeliers

CONTENTS

Throughout the previous chapters, we have analysed all the relevant elements to consider when planning, developing and implementing an urban interchange. It follows a multiagent process that we have called the City-HUB life cycle, defined in Chapter 4. This life cycle starts from the identification of requirements and needs, then continues with the validation and deployment of the interchange project in a specific location, which will have a dynamic interaction with business and activities already located in the vicinity.

The final phase of the life cycle consists of monitoring all the previous actions in a continuous communication with the different stakeholders.

The description of the different elements of the interchange, and the activities that should be allocated inside its premises, are described in previous chapters, particularly Chapter 5. They follow the logic of the two interchange dimensions, as described in Chapter 3, which include the activities (transport and services) inside the interchange and those located in the surrounding area of the city. In some cases, the interchange is expected to transform the borough, but sometimes only to change some of the shops, jobs or other activities in the vicinity of the interchange, while in most cases it has a clear local impact.

Finally, all the activities at the interchange or in the surrounding area depend on how travellers perceive their added value for their daily trips and associated activities. Therefore, we need to understand travellers' perceptions and needs in order to provide high-quality services associated with the interchange activities. These issues have been addressed in Chapter 6.

This chapter explains the logic of the City-HUB model based on the description of the key relationships among the elements of the process that have been described in previous chapters. Figure 7.1 shows all these elements and interrelations along the City-HUB life cycle. However, all the actions of the interchange life cycle involve some governance measures that should be performed in parallel.

Therefore, each of the phases of the life cycle has its correspondent part in the governance stream, as shown in Figure 7.1. The last stage of the cycle consists of monitoring the interchange outputs, which serve as the basis for further improvements in response to the changing external conditions that constituted the initial goals for developing the interchange, in that way closing the double loop formed by the life cycle and governance actions.

7.1 GOVERNANCE OF THE INTERCHANGE LIFE CYCLE

The interaction between life cycle and governance actions constitutes the essence of the City-HUB model proposed in this book. This model covers all the stages of an urban transport interchange. Figure 7.1 shows all the elements and actions of the City-HUB model.

The whole City-HUB life cycle needs to be managed by the governance actions corresponding to each step of the process. There are governance actions linked to the *identification* phase, which defines policy goals and in which stakeholders could actively participate. The next phases are related to *validation and deployment*, where the governance actions consist of getting the identified stakeholders involved, with each of them playing their proper role. This interaction defines the appropriate business model for the specific needs and requirements of the interchange. Finally, governance in the *monitoring and assessment* phase aims to enhance the performance of the current interchange by producing indicative guidelines for its improvement. This will be the start of the retrofitting process to adapt the interchange to the changing needs and business activities around it.

7.2 INTERCHANGE DIMENSIONS: UNDERSTANDING ROLES AND IMPACTS

The basis of the analysis depends on a good understanding and definition of the interchange dimensions, as stated in Chapter 3.

Each interchange has to look for the fulfilment of two dimensions: transport interchange characteristics and urban integration.

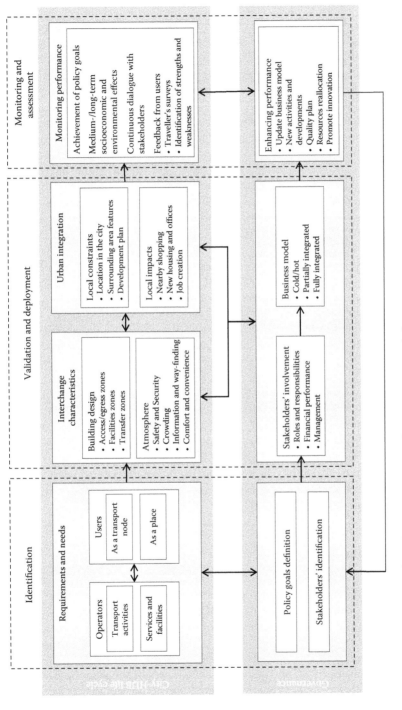

Figure 7.1 City-HUB model for urban transport interchanges: process and elements.

7.2.1 Transport interchange characteristics

The interchange plays a primary role in the city as a node where some transport activities are concentrated: transfers between different transport modes or different services of the same mode. Transport activities are the fundamental part of the interchange; otherwise, it would not be an interchange. However, linked to those transport services there are other activities that improve the transport ones, such as services linked to transport (ticket offices, luggage, etc.), but also other services that profit from the number of travellers transferring in the interchange, for example, retail/food outlets.

There is a two-way relationship between the transport and interchange services that provides positive synergies. On the one hand, the more and better-quality services placed within the interchange, the higher the perceived quality of transport services, and on the other hand, new businesses arise as the travellers who transfer inside the interchange could become customers of the different shops and services.

Everything should be properly integrated, as has been explained in detail in Chapters 5 and 6. Therefore, users have two totally connected views of the interchange: as a *transport node* and as *a place*.

The development of those views in an integrated way depends on the clear identification of the *requirements and needs* of all stakeholders that should define the appropriate *transport interchange characteristics*. These characteristics constitute the first dimension of the specific interchange that is going to be developed or renewed. Therefore, the *interchange characteristics* depend on the number and type of transport modes and the number of travellers (trip demand). They also depend on the types of services and facilities that are provided.

These considerations provide the elements for designing the *urban transport interchange* as part of the validation and deployment phases. As shown in Figure 7.1, it has two elements: *building design* and what we call *atmosphere*. The content of both elements have been explained in detail in Chapters 5 and 6.

7.2.2 Urban integration

The second dimension refers to the bidirectional interrelations between the activities inside the interchange and those located in the surrounding area.

The first direction of these interrelations is from the local area towards the interchange: what is located in the territory affects its development, that is, *local constraints*. These effects are dependent on the location of the interchange: city centre, suburbs, city border and so on. Its functionality is linked to that location, and also to the character of the surrounding area: residential, commercial, business oriented and so on. Finally, a major constraint could be the urban development plan for the area. It could define the available land for building the interchange and its auxiliary installations, the type of land uses it is possible to develop and so on. An extreme case could be if the area next to the interchange was a green area or had a hospital; therefore, some of the activities that could foster the business model of the interchange would not be developed because of use restrictions.

The second direction of interrelations is from the interchange itself towards the surrounding area. This is associated with the activities and people attracted to the interchange as a *place*. This dimension includes the *impacts on local activities*. These impacts are linked to the services inside the interchange and provide an interface that enlarges the interchange effects: more businesses will develop because the surrounding area has good access to the transport node, and travellers perceive more benefits from the interchange location. Again, there is a link between what happens inside the interchange and the activities located in the surrounding area.

These effects are considered within the urban integration box which belongs to the validation and deployment phases. How that integration could be more active and the types of relationships possible have been described in Chapter 4: interactions between City-HUB interchanges and the city.

7.3 THE APPROPRIATE BUSINESS MODEL

Chapter 3 has defined three broad types of interchanges with different requirements for the *governance* of the interchange life cycle. Once all the interchange dimensions and elements have defined the type of interchange that is necessary in each particular case, an ad hoc business model should be identified to develop and to manage it properly.

The type of business model is linked to the definition of the *policy goals* and the *identification and involvement of stakeholders* in the two first phases of the life cycle. As a result, it is possible to build the most convenient type of business model to achieve the defined policy goals in close interaction with the stakeholders.

A diverse set of definitions for the term *business model* and its components has emerged since it first appeared in scientific literature (Bellman et al. 1957). The term has generally been applied to companies aiming to gain an advantage in defined markets by conceptualising an interrelated set of decision variables in a venture strategy (Morris et al. 2005). In the case of interchanges, we propose the concept of *business model* as defined by Osterwalder (Osterwalder 2004; Osterwalder et al. 2005), but with necessary adaptations for intermodal terminals where the products are the transport activities and other services described in Chapter 5.

Osterwalder analysed the semantics of the concept of the 'business model'. On the one hand, the term 'model' means simplification of reality, and 'business' means the activity of providing goods and services involving financial, commercial or industrial aspects. On that basis, he defined a business model as an analytical tool containing a set of objects, concepts and their interrelationships used to express the business logic of a company. This tool must represent what value is provided to customers – as an interaction of supply and demand – in a simple and understandable way, and what the financial consequences are. Under these considerations, Osterwalder defined the nine objects, or concepts – *building blocks* – that a business model should be composed of. There are blocks referring to the infrastructure (offer) side and others dedicated to the customer interfaces, which include target customers and the distribution channels. Costs and revenues are included in the financial blocks. All the products and services are bundled by the value propositions of the model, which provides the strategic view of the business.

This theoretical frame of business model was adapted to the activity of interchange stations in the HERMES project (EU 7 FP 2010/2012), which aimed at exploring and thus developing models for interconnectivity. Until that moment, the application of business models in the transport sector was underdeveloped, except for a few cases of air transport terminals (Macário and Van de Voorde 2010). Based on the development of the HERMES project, a new configuration of building blocks is proposed for the case of urban transport interchanges. Figure 7.2 shows its structure, where on the left, we can find the building blocks associated with supply, and on the right, the building blocks associated with the demand of customers – or passengers. The value proposition in the central block corresponds to the possible measures for improving the services provided in the interchange stations by analysing existing supply and demand.

The left side of the first column of the table lists the identified stakeholders and how they perform their role. They offer services for the different activities located at the interchange building and manage the resources for that end – see second column. On the right-hand columns, it is first necessary to identify the users' characteristics to define the business relationships – see second right-hand column. At the bottom, the table shows the financial part of the business – costs and revenue sources. Finally, the core of the table is dedicated to identifying the actions needed to get the maximum benefits out of all the actors and assets of the interchange.

Although they are not part of the interchange business model, the activities located there also play a role in making the business model more or less robust both at present and in the future.

Each type of interchange would have a different business model. Thus, the *cold/hot* type of interchange would have a very simple one: there are very few types of travellers and transport modes that require only basic facilities. They are easy to finance and to manage. The business model becomes more complicated for the *partially integrated* type of interchanges. For *fully integrated* interchanges, the business model could be a complex challenge because of the large investments involved and the coordination of

Offer side			Demand side	
Key stakeholders	*Services* / *Resources*	*Value propositions*	*Interaction with users* / *Atmosphere*	*Users' characteristics*
Transport operators of different modes Land developers Retailing, shops Cafés, restaurants Builders, construction companies Public transport authorities City authorities Regional authorities	**Services** Transport modes Transfer among different modes Ticket selling Luggage handling services Shopping and food services **Resources** Platforms Ticket offices or machines Waiting space Information offices/screens and other devices Commercial area Area for parking (cars, cycles, buses)	Managing rapid transfer movements without congestion Exposure to shops Quality waiting areas with good ICT Coordination among public authorities Coordination with local business in the area Integration into the area and location in the city Development plan Continuous dialogue with stakeholders	**Interaction with users** Facilities for travellers Surrounding area features and activities Different information channel **Atmosphere** Safety and security Information and way-finding Comfort and convenience Entertainment	Travellers' demand for each mode Travellers' profile: • Age, gender • Frequency • Trip purpose • Income level • Disabled Non-travellers using other services
Costs			*Revenues*	
Building: construction, maintenance Energy efficiency Operation and surveillance			Fees for transport modes Renting space for retail and other services Advertising	

Source: Based on HERMES business model structure.

Figure 7.2 Business model for urban interchanges.

many different stakeholders required for putting different economic activities in place with different management priorities and payback periods.

Appendix D shows some examples of business models for different types of interchanges among the case studies of the City-HUB project.

Finally, it is worthwhile to say that any business model should not be static because any interchange must be able to respond to changing demands and technologies. It should be updated and adapted to the changing situation through the *monitoring and assessment* phase. This is a continuous process of assessment and policy adjustment in continuous dialogue with all the stakeholders: promoters, operators, public bodies, travellers' associations and so on. This process has been described in Section 4.4. It also includes feedback from travellers, who have a different way of expressing their preferences, as stated in Chapter 6. It is necessary to customise the travellers' surveys for each specific interchange to get valuable feedback from users.

7.4 IMPROVING INTERCHANGE PERFORMANCE THROUGH TIME

As has been said, the management of the interchange does not finish when the construction is completed and all the services are settled. A number of activities should be carried out in order to assure the usefulness of the interchange activities or even carry out improvements with experience. That means a continuous process of *monitoring and assessment*. This process should be coordinated by the *interchange manager* in close contact with the public authorities that promote it and the stakeholders that finance and promote the different activities: transport operators, cafés, retailers and so on.

The first task is to check the *achievement* of the initial *policy goals* as defined in the *identification* phase. It is then necessary to undertake an analysis to assess whether the goals are still all valid. Normally, this will be not the case, as new requirements and needs could appear with time, demanding new activities or the modification of initial ones.

It is then necessary to evaluate the *urban integration* aspects of the interchange; checking if the foreseen local impacts are as expected in the medium–long term. The environmental impacts should also be assessed to identify which activities could produce negative impacts. This is not a simple process and to that end it is necessary to get the stakeholders' opinions through ad hoc surveys and interviews. Those opinions should be complemented with data collected in the area: for example, the number of new housing and offices, shops, changes in the number and type of jobs. Environmental quality standards should also be measured: for example, noise, pollution and deterioration of green areas.

Users' opinions are also very important in enhancing an interchange. Chapter 6 has shown how it is possible to know what the travellers'

Table 7.1 City-HUB model elements table

Life cycle phase	Element	Where in the book	
		Chapter	Section
I: Identification	Transport operators	2	2.3.2
		3	3.2.1
		4	4.1.2
	Transport users	2	2.3.3
		3	3.2.1
		4	4.1.2
		5	5.1.4
		6	6.6
	Policy goals	1	1.1, 1.2
		2	2.1, 2.2, 2.3.1, 2.4.2
		4	4.1.1
		8	8.2, 8.5.1
	Stakeholder identification	2	2.3
		4	4.1.2
		8	8.2, 8.5.1
II and III: Validation and Deployment	Building design	2	2.1, 2.3.3, 2.4.2
		3	3.2.3, 3.3
		5	All
		6	6.2
		8	8.3.1, 8.5.4
	Atmosphere	2	2.1, 2.3.3, 2.4.2, 2.5.1
		5	All
		6	6.2, 6.7, 6.8
		8	8.3.1, 8.5.4, 8.5.6, 8.5.7
	Local constraints	3	3.2.2, 3.3
		4	4.3.2, 4.3.3
		7	7.2
	Local impacts	3	3.2.4
		4	4.3.2, 4.3.3, 4.4.2
		7	7.2
	Stakeholders involvement	3	3.2.5
		4	4.1.2, 4.2, 4.3.1
		8	8.3.2, 8.5.2
	Business model	4	4.2.1, 4.2.2
		7	7.3
		8	8.3.2, 8.5.2
IV: Monitoring and Assessment	Achievement of policy goals	4	4.4.1
		8	8.4
	Medium–long-term socioeconomic and environmental effects	4	4.4.2
		8	8.4, 8.5.1
	Continuous dialogue with stakeholders	3	3.2.5
		4	4.1.2
	Feedback from users	4	4.1.2, 4.4.3
		6	All
		8	8.5.3
	Enhancing performance	4	4.4
		7	7.4
		8	8.5

expectations are, and the value of the perceived quality of the services when using the interchange: that is, the transfer among modes, retail and other services which minimises the penalty of changing from one mode to another.

The results of this assessment phase are really the basis for assuring the quality of the interchange along its lifetime. Based on these assessments, the business model could be updated, incorporating new stakeholders and modifying the priorities as necessary. That means considering new activities inside the interchange or in the area, along with new development plans. The *quality plan* should report what has already been achieved and enhance it with new activities. All in all, the interchange is a good place to promote innovation and new solutions. Its services are very much linked to urban transport overall. The large number of users and the high demand for changing information inside the interchange – the coordination of the different transport modes and services – convert it into a living laboratory to test and develop new ICT-based solutions.

7.5 MAPPING THE CITY-HUB MODEL

In order to facilitate the location of the different elements of the City-HUB model in this book, the Table 7.1 lists where the reader can find more information about each of the boxes included in Figure 7.1.

Part III

Benchmarking the City-HUB model

Chapter 8

Cross-cases analysis

European interchanges

Petter Christiansen, Jardar Andersen, Gábor Albert,
Ádám Pusztai, Álmos Virág and Riccardo Poppeliers

CONTENTS

8.1 CASE STUDIES APPROACH

A standardised approach for data collection was necessary in order to benchmark the City-HUB model. This benchmark study of European urban transport interchanges is based on two complementary approaches: the validation of the City-HUB model through a set of checklists and good practices for resolving them, and monitoring and assessing the interchange. The checklist set was implemented on six selected validation case studies, which were:

1. Gare Lille Flandres/Europe (France)
2. Utrecht Centraal (the Netherlands)
3. Oslo bus terminal Vaterland (Norway)
4. Paseo de Gracia, Barcelona (Spain)
5. Prague terminus, Dejvická (Czech Republic)
6. Intermodal terminal of Miskolc (Hungary) (under planning)

The six validation cases represent diverse characteristics in terms of modes covered, size of interchange, number of passengers handled, location in the urban area, time since construction and stakeholders involved (see Appendix I). Yin (2009) points out that substantial analytical benefit arises from using comparative studies.

The comprehensive checklist set can be used to evaluate interchanges and identify areas where improvements are needed. The elaboration of the checklists is based on the analysis of the pilot case studies and interviews with practitioners of a wider sample of the European urban interchanges (i.e. 21 urban interchanges) presented in Table 1.1 in Chapter 1 and in Appendix II, and the treatment of qualitative and quantitative data from the user surveys realised in five pilot case studies (see Table 1.1, Chapter 6 and Appendix III). The checklists address:

- Identification of policy goals, stakeholders and stakes.
- Validation and deployment: checking key interchange aspects (i.e. travel demand, dominant transport modes, services and facilities, location in the city and local impacts), and properties (interchange design and accessibility, coordination of intermodal transport, information and way-finding, facilities and services, and safety, security and environment).
- Governance issues: business models.

- Monitoring and assessment (achievement of policy goals, environmental effects and user feedback).

Each partner responsible for a validation case study was asked to assess the interchange by using various checklists. The stakeholders contacted in the case studies were asked to complete the checklists and assess the interchange status for each indicator. The checklist used a Likert scale from 1–5, where 1 is the lowest grade and 5 is the highest. By using this tool, the actors responsible for analysed interchanges can identify areas that work particularly well and areas that need improvements. Consequently, the checklist can be understood as a tool for improving current interchanges. Stakeholders can use this in order to answer the most basic questions, such as What is the current state of the interchange and what are the most important aspects that need to be improved in order to offer travellers (i.e. users) good facilities for transferring between modes?

For new interchanges, the actors involved can use the checklist to validate that they have included all the necessary aspects for designing a successful interchange. The various aspects in the checklist have also been based on extensive literature reviews and pilot case studies analysis. The checklist can thus make the actors aware of important interchange qualities that they have not sufficiently considered.

It is important to note that the scores for the different interchange properties depend on the judgement of those who have completed the checklists. We have attempted to reduce this problem by making the responsible partners explain the results obtained through their own case study.

For the planned new intermodal terminal (i.e. Miskolc in Hungary), scores could not be given; in that case, the checklists were used to assess whether the different aspects were considered or not in the planning.

The second tool for interchange benchmarking is related to the good practice identification that is based on the results of interviews (questionnaires reproduced in Appendix II), the pilot case studies analysis, and sometimes includes the results of the checklist implemented in the validation case studies.

The results of the implementation of the checklists on the validation case studies are discussed in Sections 8.1 through 8.4, which are organised by the four categories of information covered by the checklists. The scores for each case study and indicator are summarised in the following tables and average values are calculated. At the same time, a set of good practices has been identified through the case studies.

8.2 IDENTIFICATION OF POLICY GOALS, STAKEHOLDERS AND STAKES

This category involves aspects which should be considered prior to or during the planning stage for upgrading or building new interchanges. It is less relevant for already existing interchanges that have no plans for implementing significant changes in the design or operation of the interchange. Therefore, some of the case studies have not filled out the checklist for these issues.

Stakeholders need to assess whether the planning of the interchange corresponds with overall (policy) goals at the national, regional and/or local level. Interchanges can have a crucial impact on land use in the most central areas of cities or regions (Di Ciommo 2004) and it is necessary to take into account operators' future perspectives for development. It is thus imperative that the responsible actors ensure that the interchange development is in line with the overall goals of transport and land use development in the region.

The checklist also includes the degree of stakeholder involvement. The project has identified that many actors can potentially be involved in the planning and operation of interchanges. With many actors being involved, a complex decision-making process is called for, since the possibility of contradicting interests between the stakeholders can be increased. It is also well documented that stakeholders can veto or stall development and such possibilities increase with multiple stakeholders (Pressman and Wildavsky 1973). However, including several stakeholders also means that the decision makers receive different perspectives on interchange development, different information and several ideas on how to pursue further progress. The aim is to make the decision-making process more transparent; to gather more input on which to base decisions; and to create support for the decisions that are made.

It is also important to identify stakeholders and their interests at an early stage in the process. This is necessary in order to achieve outcomes that are attractive for all parties. The leading actors need to combine and gather all of the various information and perspectives in order to manage the complexity of the process. Involving stakeholders at a late stage can prove to be inefficient. Possible scenarios include delays due to unresolved issues, amendments of development plans or the need for further information or studies in order to ensure that all of the stakeholders' issues are resolved. However, it is also a prerequisite that the stakeholders have clarified and anchored their positions within their own organisation. Otherwise, the process might halt due to changing interests or the introduction of new aspects that have not been discussed at an earlier stage.

Regardless of the different categorisations, the responsible actors need to define the functions and logistics at the interchange (see Chapter 3). The functions and logistics set a framework for the overall development and therefore need to be clarified at an early stage. An essential question is whether modes of transport are ending their route at the particular interchange or whether the modes of transport only stop for a short time. This has consequences, for example, for the development of platforms. Other questions and aspects are how (other) authorities and operators have planned for the future route network, investments and land use developments in adjacent areas at the interchange and in a regional perspective. This is particularly relevant for the future demand of passengers and routes at the interchange. Small interchanges may have more simple transport-related functions, but the larger ones are also dependent on retail and commercial opportunities.

8.2.1 Results related to policy goals, stakeholders and stakes

Table 8.1 summarises the results in terms of policy goal, stakeholders and stakes for each interchange validation case. The table shows great variations between the different interchanges. For the planned new intermodal terminal of Miskolc, scores could not be given; in that case, the checklists

Table 8.1 Checklist related to policy goals, stakeholders and stakes

Checklist	Lille	Utrecht	Oslo	Barcelona	Prague	Average
Identify policy goals/needs for actions						
Identify policy goals and needs for actions (external and internal)	N/A	N/A	N/A	2	3	2.5
Identify stakeholders and stakes						
Stakeholder strategy outlining the engagement process	N/A	N/A	N/A	2	2	2
Engage hard-to-reach groups such as disabled people	5	N/A	N/A	4	5	4.7
Follow the seven key principles for consultation: integrity, visibility, accessibility, transparency, disclosure, fair interpretation, publication	5	N/A	N/A	3	4	4
The stakeholders' roles are clear and anchored within the organisation (clear responsibilities)	5	N/A	N/A	4	4	4.3
Transport services are of a significant size that provide good accessibility and connections between modes	5	N/A	N/A	2	3	3.3
Define functions and logistics of the interchange						
Definition of the interchange clearly laid out	5	N/A	3	3	4	3.8
The logistics, site, connectivity with selected transport modes, volume of passengers suitable for the interchange	N/A	N/A	3	5	2	3.3

were used to assess whether the different aspects were considered or not in the planning. This is still an auto-evaluation that we need to mitigate because it could depend on the optimistic attitude of the interchange stakeholder who is evaluating her/his case study. Interviewed stakeholders from Oslo, Utrecht and, to a lesser degree, Lille, preferred to not evaluate some of the checklist criteria (i.e. policy goals and stakeholders' involvement).

The Lille case study seems to have the best marks, with some items that have not been evaluated. The responsible partners give themselves a rating of five on the majority of issues, while Paseo de Gracia in Barcelona receives the lowest average score in the auto-evaluation exercise for the checklist related to policy goals, stakeholders and stakes. According to the interviews, the Lille interchange identified the needs of the operators and clients before refurbishment of the interchange. They also consulted the reduced mobility persons' association, and the Lille Flandres has discussion groups for the operation. However, these issues are most relevant in cases of planning new interchanges. A lack of information about the stakeholders' involvement in Utrecht and Oslo interchanges highlights the lack of good practices in considering the interchange infrastructure as a transport infrastructure for improving its use and management. An important effort should be made to better define the appropriate and comprehensive business model for each interchange.

8.3 VALIDATION AND DEPLOYMENT

When considering improving an interchange, it is a necessity that responsible stakeholders identify problems that require actions. In order to do so, the checklist can be a helpful tool, since it defines the most important interchange properties. It defines five key factors that are important to ensure good facilities for travellers: interchange design and accessibility; coordination of intermodal transport; information and way-finding; facilities and services; and safety, security and environment. All these aspects can contribute to reducing the disbenefits associated with switching modes (interchange penalty) (see Chapter 5).

Interchange design and accessibility encompass aspects related to good links within and outside the interchange. For the outside areas, it is important to have good accessibility. That means that the infrastructure should have links with external facilities, with direct, convenient and clearly identified routes to stations (see Chapter 5). The routes should also have easy and unobstructed access. This is important for both frequent and less frequent travellers. Good accessibility with unobstructed access supports reduction of total travel time and can influence the use of soft modes such as walking and cycling to the interchange.

Once inside the interchange, travellers need safe, step-free and direct routes between modes that minimise conflicts with other travellers and vehicles and allow access for all users. This will shorten transfer times. However, it is not only important to have good routes and links between modes. The interchange should also be a pleasant place to be in. This means that waiting areas should be well lit, heated and ventilated, noise levels should be comfortable and cleaning and maintenance should take place without affecting the day-to-day operations, as shown in Chapter 5. These aspects are aimed at increasing the comfort level for passengers. The attractiveness of the interchange can be reduced if travellers find it to be unpleasant. Travellers do not generally enjoy waiting. Internet access and computer facilities can also improve productivity, since travellers can spend their waiting time on activities such as reading and checking emails (Lyons and Urry 2005).

Coordination of intermodal transport is another aspect that supports the use of public transport. Interchanges often involve various modes of transport and operators decide how the ticket systems are organised. However, the interchange can potentially set requirements as to where and when tickets are sold for travellers at the interchange. This system should take into account the needs of different users and be organised in a manner that reduces any barriers connected to changing between modes. The ticket offices or automated ticket machines should therefore have good capacity and a suitable location in order to make the ticket purchase as easy as possible. Information can be written in different languages to ease the purchase operation, particularly at international interchanges.

Facilities and services are factors that can increase the attractiveness of the interchange. Firstly, high-quality facilities and a wide range of retail options can improve waiting times. Comfort is a key driver of behaviour (Hine and Scott 2000). Secondly, such services can make it possible for travellers to combine different purposes. The possibilities for using public transport can increase since travellers can shop or buy groceries in combination with, for instance, work trips.

Travellers and operators also need to feel safe and secure at the interchange (Coccia et al. 1999). In order to facilitate this, the interchange needs to keep flows of vehicles and passengers separate to reduce the possibility of collisions and accidents. They also need to feel secure by having staff available and/or having CCTV inside and in the areas surrounding the interchange. The design should, in addition, take into account that the access points should feel safe and secure at all times of day.

The last point is to have high-quality information and way-finding. For infrequent travellers especially, it is of vital importance that they can easily find information across modes and easily orientate themselves within the interchange. This means that the interchange should offer integrated real-time information across modes and have simple and clear signage for all users. Such aspects can be facilitated by having a self-explanatory design that is easy to follow and thus minimise the number of signs required. New technologies have also made it possible to offer information, through mobile phones and Bluetooth devices. This can particularly relevant for people with reduced mobility as shown in Chapter 5.

8.3.1 Validation and deployment: results

We have shown the average score for each element in Table 8.2. In addition, we have an average score for each interchange. By doing so, we can analyse whether there are particular areas that function better than others. We can then show whether there are common challenges among the interchanges. We can also make comparisons between the interchanges and show whether some are rated considerably higher.

Comparing the results of the travellers' satisfaction questionnaire related to key elements of interchanges (see Chapter 6, Table 6.5), it seems that interchanges' practitioners are more optimistic about safety and security elements. Actually, users scored around three or less than three. Therefore, a real involvement of users in the monitoring and assessment of the interchange infrastructure will be needed to identify the strengths and weaknesses of the managing and maintenance of the interchange.

The table illustrates the average values for each of the key interchange properties. In general, the case studies are functioning best with regard to the coordination of intermodal transport. In particular, the interchanges have high scores for the capacity and location of ticket offices, how clear ticket options are and the ease of purchasing tickets for all users. Three out of five interchanges, however, have low scores on integrated ticketing. For instance, in Utrecht, it is not possible to buy tickets for public transport in the Utrecht region. The same applies for Oslo Vaterland.

Safety and security ranks second in terms of key interchange elements. All interchanges use CCTV and the flows of vehicles and passengers are in general separate. The element that receives the lowest value is the use of automatic passenger counting systems to analyse passenger flow in and out. Such an element can be understood as a tool for improving safety and security and can therefore be less relevant compared to the other aspects.

Information and way-finding, as well as facilities and services, receive about the same values. Both of these aspects are essential qualities for travellers. The case studies have the best average scores for the elements 'Illuminated routes should be evenly lit, and any sudden changes in lighting levels, glare, dark spots or pooling should be avoided' and 'lighting from retail and other commercial outlets should not detract from the lighting provided for the improvement of legibility and accessibility'. The main areas with room for improvements are 'way-finding should be provided in the area surrounding the interchange' and 'the interchange offers integrated information across modes in real time'. For instance, in Lille Flandres there is no clear direction information on finding the Lille Europe

Table 8.2 Checklist on interchange key elements

Checklist	Lille	Utrecht	Oslo	Barcelona	Prague	Average
Interchange design and accessibility						
Waiting areas are well lit, heated or ventilated	3.5	2.5	4	4		3.5
Noise level is comfortable	4	2.5	5	4	1	3.3
Cleaning and maintenance take place without affecting the day-to-day operations	5	5	5	5	4	4.8
Routes and facilities are of a high quality and appropriate for the type of interchange	5	3	4	3	3	3.6
Safe, step-free, direct routes for travellers between modes that minimise conflicts with other travellers and vehicles and allow access for all users	3.5	N/A	3	4	2	3.1
Walking distance, convenience and legibility are of high quality for the type of interchange	4	3	3	1	3	2.8
Good links with external facilities meaning that there are direct, convenient and clearly identified routes to/from stations, with easy and unobstructed access	3.5	3	3	4	4	3.5
Tactile surfaces for guidance and information, personal assistance and more advanced guidance solutions	4	4	3	1	2.5	2.9
Internet access (Wi-Fi) and computer facilities are offered	5	3	4	1		3.3
Pedestrian and cycling modes have the highest priority followed by public transport modes	2		2		2	2
Good cycling infrastructure in and around the facility	5	3	1	4	1	2.8
The evacuation routes also have waiting and refuge areas for people with reduced mobility	5	1	4	3	2	3
Low emissions from vehicles indoor	5	N/A	5	5		5
Coordination of intermodal transport						
The interchange has an integrated ticketing system	5	1	2	5	2.5	3.1
Good capacity and location of ticket offices	5	3.5	5	5	4	4.5

(Continued)

Table 8.2 (Continued) Checklist on interchange key elements

Checklist	Lille	Utrecht	Oslo	Barcelona	Prague	Average
Ticket options (such as zones and validity) are clear and purchasing the tickets is easy for all users	5	2	5	5	4	4.2
Locations for picking up and setting down for taxis are clear to the user and also support the layout of other transport modes at the interchange, not impeding local traffic flow	5	3	3	3	4	3.6
Facilities and services						
High-quality facilities and wide range of retail options	3.5	4	2	3	3	3.1
Spatial organisation promotes good quality of user movement.	3	3	3	3	4	3.2
The needs of the different users, both arriving and departing, as well as special needs should be considered	4	3.5	N/A	4	4	3.1
The interchange zone (the public area serving the facility itself) is included in the implantation of the facility into the city or town scape	3.5	3	1	4		2.8
Safety and security						
Flows of vehicles and passengers are separate	5	4	3	5	3.5	4.1
Automatic passenger counting systems (e.g. infrared sensors) are used to analyse the passenger flow in and out	2	N/A	1	4	4	2.7
Use of CCTV inside and in areas surrounding the interchange	5	4	4	4	4	4.2
Possibility for human interaction at the interchange	5	4	4	5	2	4
An emergency management plan that has been agreed amongst stakeholders	5	3		3	4	3.7
All access points should feel safe and secure at all times of day	4	N/A	4	4	4	3.7

Information and way-finding

Criteria						
The interchange offers integrated information across modes in real time	4.5	4	1	2	3	2.9
Simple and clear signing for all users identifying transport information; different local facilities; temporary information; and safety and warning information	5	4	2	4	4	3.8
The interchange has a self-explanatory design that is simple to follow and thus minimises the number of signs required	3	4	4	4	4	3.8
The interchange has a way-finding plan	5	4	3	2	2	3.2
Use landscaping and lighting to complement signage and increase user legibility. A legible environment ensures the easy and seamless navigation and movement of users	5	N/A	3	2		3.3
Provision of clear connections to existing routes, facilities and services for allowing users to move around the interchange under several alternatives (i.e. permeability)	4	N/A	3	4	4	3.7
Information supported through various technologies, such as audio and visual displays, mobile telephones and Bluetooth devices. The designing of signage should include consideration of cultural differences, language differences, cognitive impairments, visual impairments and mobility impairments	4	3	2	3	N/A	3
Way-finding should be provided in the area surrounding the interchange	2	3.5	1	3	3	2.5
Illuminated routes should be evenly lit, and any sudden changes in lighting levels, glare, dark spots or pooling should be avoided	4.5	4	4	4	4	4.1
Lighting from retail and other commercial outlets should not detract from the lighting provided for the improvement of legibility and accessibility	2.5	4	5	5	N/A	4.1
Seamless connection between the interchange and the surrounding areas and any external destinations	4	3.5	1	5	2	3.1
Personal information through e.g. near field communication based on smartphones for access to service timetables, information about planned service outages, etc.	5	3.5	1	4	N/A	3.3

interchange, and in Lille Europe it is not obvious how to reach the tram station.

The final aspect is interchange design and accessibility. On average, this receives the lowest score. The most challenging areas are walking distance and tactile surfaces, as well as cycling infrastructure in and around the facility. Barcelona specifically has a low score with regards to walking distance, convenience and legibility. The distances between the modes of transport at this interchange exceed 400 m.

Universal requirements are also necessary in order to provide a transport system that includes the needs of those with reduced mobility. Both Barcelona's Paseo de Gracia and Prague's Dejvická have poor conditions. In the latter, there is good guidance in the refurbished part of the interchange, but no guidance and poor information for disabled users in the old section of the interchange. This is a problem that has been identified in several of the cases. There is great variation in standards within the interchanges themselves.

It is also possible to compare the average values for each interchange in Table 8.2. Lille receives the best scores in almost all checklist categories, followed by the Barcelona case study, while Prague, Oslo and Utrecht are scored lower with a normal score value of around three out of five. Lille interchange is also among the most recently upgraded, where the involved practitioners had the opportunity to be aware about various key elements of an interchange and auto-evaluate themselves.

8.3.2 Validation and deployment: managing an interchange

This part of the checklist captures the business and management side of interchanges. This includes mapping out opportunities to link up with the private sector and opportunities to combine space with commercial interests, as well as providing an investment and maintenance plan that includes identifying opportunities for external revenue collection.

The service level is a critical determinant for a successful interchange. In principle, a public stakeholder, the private sector or a combination can manage interchanges. However, according to the case studies in the City-HUB project, a public actor manages the majority of the interchanges. The case studies also documented several cases where the public sector did not pay attention to the financial viability of the interchange construction and maintenance.

Depending on the local context, interchanges can generate high economic impacts and revenues (Banister and Berechman 2001). They can also be an important driver for economic development and land use development in areas adjacent to the interchange. Consequently, the authorities could utilise business links and use the private sector to develop cost-efficient ways of constructing or managing interchanges.

One of the main purposes for mapping out opportunities to link up with the private sector is that the private sector can contribute extra resources in the construction phase. Extra resources can be used to increase the financial sustainability of the project and facilitate the integration of businesses, shopping and/or housing in the interchange development. This can be especially important in instances where authorities lack finances or when the organisational set-up mainly caters for aspects such as accessibility and comfort, and does not have incentives for taking into account other perspectives that also could generate further income. Linking up with the private sector can also encourage the local authorities to take into account wider perspectives.

Another main purpose for mapping out opportunities to link up with the private sector is the necessity to analyse the financial performance of the operational phase. By taking into consideration financial sustainability, the responsible stakeholders need to have a separate breakdown of future costs and revenues (Di Ciommo et al. 2009). This could provide incentives for linking up with the private sector in order to receive further income.

Securing complementary policies and a development plan can be essential elements in achieving a well-functioning interchange, since an interchange is not only dependent on its own facilities, but also on the development of adjacent areas (Heddebout and Palmer 2014; Banister and Berechman 2001).

First, securing good accessibility for modes such as bus, car, walking and cycling is vital for making public transport attractive to use. The case studies have revealed that although an interchange is well functioning within the areas of its own responsibilities (distances between modes, facilities for retail and services), there are challenges connected to congestion, a lack of cycle paths or a lack of accessibility for walking that increase the travel time for buses and cars. Securing complementary policies is thus necessary in order to improve the attractiveness of the interchange. It is well documented that appropriate infrastructure, such as cycle paths or dedicated bus lanes, improves travel time and thus increases the use of such modes. Therefore, it is important that the responsible partner, at an early stage, attempts to secure complementary policies with the relevant municipal or other authorities. This is essential in order to maximise the effects of the investments in building or improving an interchange.

Secondly, there are examples of interchanges that constitute a vital part of an overall plan for larger refurbishments or developments. Urban development policies are then linked to interchange improvement in order to advance urban functions, service functions and economic development. In general, the urban development plan aims to provide new shopping facilities, new housing and new offices that in turn are necessary components for making an interchange more attractive.

This illustrates the necessity of political willingness to implement complementary policies in order to provide a better environment, to boost the

Table 8.3 Indicator values related to validation and business model assessment

Checklist	Lille	Utrecht	Oslo	Barcelona	Prague	Average
Business model						
The interchange has mapped out opportunities to combine space with commercial interests	5	N/A	I	2	2	2.7
Financial plan covering the investment, maintenance and operating costs and revenue flows	5	N/A	2	I	2	2.7
The interchange has a business model	5	N/A	3	I	I	3
Retailers complement one another and serve the vision of the attractiveness of the interchange	5	N/A	I	3	I	3
Secures revenues to finance future upgrades	5	N/A	3	2	I	3.3
Political willingness to implement complementary policies	5	N/A	2	2	I	3
A development plan that coordinates land use and city integration	5	N/A		3	2	4
High-quality labour force	5	N/A	5	I	I	3.7

investment in transport and obtain economic development. These complementary policies can be to implement interchanges as part of an overall larger integrated policy and/or plan aiming to (re)develop linked economic activities and urban function (re)development. Moreover, Banister and Berechman (2001) conclude that 'policy design also has a crucial role in influencing and strengthening the potential impact of transport infrastructure investment on local economic development.'

The results in Table 8.3 illustrate a clear distinction between Lille on the one hand and Oslo Vaterland/Barcelona Paseo de Gracia on the other hand.* According to the Lille case study, new offices have been planned along with a wider range of retail opportunities. About 40,000 persons enter the Lille Flandres train station daily only to use the services or shops or to use it as a crossing path between two parts of the city.

This is in contrast to the Oslo Vaterland case study. The structure of the business model has been criticised for being unclear. An operating company has recently initiated an ongoing lawsuit against Akershus Public Terminals

---

* For the Utrecht case study, it was not possible to receive information from the interviews regarding validation and the business model.

and the county. The company argues that over many years they have paid too large a share of the costs of operating the terminal. The conflict concerns Ruter not paying departure charges, how these are calculated and how the incomes are distributed between Akershus Public Terminals and Vaterland Bus Terminal AS. The business model and responsibilities are not clear amongst all stakeholders, which has been criticised in a consultant report (Analyse & Strategi 2012). The interviewees also explained that they are not interested in facilitating commercial activities at the interchange. They want to focus on the public transport services and consider the existing kiosks/cafés to be sufficient. Those responsible interviewed in Utrecht preferred to not score the aspects related to the business model.

8.4 MONITORING AND ASSESSMENT

Checklists include aspects that highlight the necessity of receiving feedback from users and evaluating performance at the interchange. Feedback can be institutionalised by having regular meetings with operators and/or users at the interchange. It can also be done by conducting regular surveys aimed at users, as shown in Chapter 6. By doing so, the interchange can get constant feedback on areas that function well or poorly. The responsible actors then have the possibility to evaluate how the interchange is functioning and implement possible solutions to improve the status. The checklist tool jointly with the users' survey could be an easy way to monitor and assess the interchange. The results of the checklists and the analysis of the survey should be public and accessible to all stakeholders in order to improve the organisation, maintenance and managing of the interchange.

8.4.1 Results related to environmental effects and monitoring

Energy efficiency and environmental effects have gained increased attention in recent years. Energy-efficient buildings can have significant impacts on the environment and can reduce operating costs for the owners of the building. The checklist is only meant to capture this issue by making the relevant stakeholders aware of the importance of energy efficiency. For instance, there are a number of different assessment tools available that could be used to make sure that the interchange satisfies requirements for environmental aspects. BREEAM or CEEQUAL are two examples.

The results in Table 8.4 suggest that actors from Lille state that they have very good standards relating to energy efficiency.* For instance, they have an objective to make a 20% energy reduction by 2022. They have

---

* The indicators have missing values for Utrecht, Oslo and Prague. This is because the actors found it difficult to rate the interchange for these categories.

Table 8.4 Checklist about environmental effects and monitoring

Checklist	Lille	Utrecht	Oslo	Barcelona	Prague	Average
Deployment						
Interchange management plan that stated the roles and responsibilities of all those involved in the interchange	Yes	N/A	N/A	2	N/A	2
Low and zero carbon energy construction criteria: photovoltaic solar panels, low energy ventilation and heat recovery, combined heat and power systems, ground coupling, solar water heating	5	N/A	N/A	2	N/A	2
Energy efficiency and low and zero carbon energy	5	N/A	N/A	1	N/A	1
Monitoring						
The interchange monitors and evaluates performance and have quality indicators	Yes	N/A	2	1	N/A	1.5
Feedback from users' perception	Yes	N/A	2	2	N/A	2

therefore started to implement several measures to reduce energy consumption, particularly from lightning. Lille also undertakes regular enquiries connected to the level of satisfaction from users. This is in contrast to Oslo and Barcelona, which have low scores on this issue. Oslo Vaterland does have meetings with operators on an irregular basis. They have also conducted surveys in order to receive feedback from travellers. However, the most recent survey is over ten years old.

The City-HUB validation case studies have tested the City-HUB life cycle process in six real-life contexts across Europe. In this chapter, we discuss the main feedback to the final version of the City-HUB model. The next section summarises the good practices that could be extracted from each validation case study.

8.5 GOOD PRACTICES

The Oxford English Dictionary has no definition of 'good practices', but defines 'best practice' as: 'commercial or professional procedures that are accepted or prescribed as being correct or most effective'.

Best practice can therefore be understood as a set of guidelines; a process, practice, design or system that represents a recommended or practical action aimed towards reaching a goal or a particular desired outcome. The best practice can thus illustrate and highlight the required standards or quality for an interchange and address a solution to a particular feature. Moreover, the best practice can be used in benchmarking, where the responsible stakeholders conduct a self-assessment on the standards in relation to the best practice. This could in turn inspire the responsible stakeholders to implement the suggested improvements.

There are several examples of using best practices in the field of transport. Transport for London has, for instance, developed such a tool. In their 'Interchange Best Practice Guidelines 2009', they offer a set of guidelines that encompass four themes for interchange development. The KITE project has also developed a catalogue of Best-Practice Implementation Examples for interchanges across Europe (Grafl et al. 2008).

However, a 'best practice' needs to be defined in order to reduce a subjective or random selection of practices. Ideally, the best practice should be documented to have significant effects that are comparably better than other measures. If not, it is a challenging task to identify and explain the best practice.

In the City-HUB project, we have described and analysed different practices for 11 case studies across Europe. This input was used to provide a set of recommended guidelines and processes for developing interchanges. It has not been possible to make quantitative targets for determining if a practice is good or not; the selection has been made from judgments based on what is perceived as good by users, operators and researchers. Consequently, we mean that it is more fruitful to use the term 'good practice'. By doing so, we acknowledge that other practices might be just as good as the ones identified in the City-HUB project. In addition, such an approach is more flexible since it allows for multiple good practices.

The good practices are based on elements that were already identified as important for a successful interchange from the perspectives of users, operators and owners. The 'empirical universe' is the 11 case studies, in a few cases supported by information from additional interchanges that have been studied in the project. It is therefore necessary to emphasise that the good practice is only based on limited empirical material. We have not included (better) examples from interchanges that are not a part of this project. This means that we run the risk of missing practices that are better than the ones identified in this project. It also means that some of the categories do not have a description of good practice since the case studies did not meet the criteria for being highlighted in this chapter.

The explanations for each good practice follow the same structure. Each good practice has a description of the problem that it is meant to amend. Then, the good practice is explained and the possible effects are described. At the end, the transferability aspect is discussed since we explain for what kind of interchanges the good practice is suitable for.

8.5.1 Identification

8.5.1.1 Assess interchange status

The earlier chapters have shown that there is a wide range of factors that are important for achieving a successful interchange. For travellers, operators and owners it is essential that the services and facilities are of such a quality that the use of public transport is promoted. At the same time, the case studies conducted in the City-HUB project have revealed that the interchange status varies greatly. Some of the interchanges are new with considerably better standards compared to the ones that have been in place for decades. However, for each of the studied interchanges, there are aspects that could be improved to offer better facilities and experiences for users at the interchange. For instance, the analysis of the validation case studies in Sections 8.1 through 8.4 showed that for the majority of the interchanges, there is room for improvements connected to, for example, signing and providing information to the travellers. These two elements are key for any interchange regardless of size. The case studies also documented that it is not common for the interchange managers to conduct assessments of the current status of the interchange.

- *Description of the good practice:* The responsible stakeholders involved at an interchange can receive valuable feedback if they assess the interchange status, especially when considering improving an interchange. It is necessary that the responsible stakeholders assess the interchange status since it can help them to identify important characteristics that otherwise would not have been detected. By identifying problem areas, the interchange managers have the necessary information for implementing upgrades to improve the facilities and services at the interchange.

  From the case studies in Barcelona, Prague Dejvická, Miskolc and Lille, it was stated that the future development of the interchanges will be affected by the areas of improvement identified by the use of the checklist. In Dejvická, the interchange manager and operator declared that it is likely that the checklist will support the planning process by reflecting missing aspects. In Lille, the interchange manager stated that the checklist has identified areas of improvement and it is likely they will make changes connected to signalling and information. The interviews from Oslo showed that it can be challenging to look at the interchange with 'fresh eyes'. The checklist tool can overcome such a problem by highlighting aspects important for an interchange. This feedback illustrated that the checklist can function as a tool for improvements and shows the relevance of the model.

- *For whom is it good practice?* Identifying areas of improvement is vital for any type of interchange. The interchange can increase the attractiveness of public transport and receive more passengers if the standards at the interchange and the public transport system are of sufficient quality. By improving the facilities and accessibility for travellers, the interchange can generate more income, which in turn can finance further improvements at the interchange or reduce the need for public subsidies. In addition, it can benefit operators if the improvements lead to more passengers.

- *For what kind of interchanges is the good practice suitable?* The interchange status must be assessed for all types of interchanges, but this is most important for the largest interchanges located in central areas and in interchanges with many passengers. Smaller interchanges located in less central areas are typically less complicated. The need to assess the interchange status is thus most important in cases where there are many users. That does not mean that smaller interchanges should not conduct assessments.

8.5.1.2 Identify policy goals/needs for action

The intention of a good interchange will generally be to improve the quality of public transport services and support seamless door-to-door travel. But nowadays an interchange is more than just a simple node in a network – it has many elements. The interchange contributes to lowering CO_2 emissions through allowing a modal shift of mechanised trips from private vehicles (specially from cars) to intermodal trips where the good quality of interchanges play a key role. Research literature shows that the benefits of urban interchanges relate to time savings, better use of waiting times, urban integration and improved operational business models (Di Ciommo 2002). Besides accessibility improvements, management and innovation, an efficient use of interchanges should also be depicted.

- *Description of the good practice:* The case of Den Bosch (the Netherlands) shows that from the very start, the municipality as a manager should clearly set goals and needs for actions, and identify and involve stakeholders. An important aspect together with the process of setting clear roles and responsibilities is to take a clear decision about a financial plan. When the city has clear roles and responsibilities and realises that another stakeholder will not fulfil the goals of the municipality, the municipality should come in and take the lead in the process (and bring in funding whenever it is considered to be a priority).
- *For whom is it good practice?* Identifying areas of policy realisation improvement is vital for any user of the City-HUB model. The interchange process can be effectively and efficiently improved by having clear goals per stakeholder and – accordingly – formulated actions and behaviour (including financial responsibilities). This step is the start of the governing process of the City-HUB model.

- *For what kind of interchanges is the good practice suitable?* Interchanges are more likely to have an impact on the local economy and land use when integrated policies are implemented in cooperation with relevant stakeholders. The existence of an integrated development plan is therefore important to induce economic development.

8.5.1.3 Identify stakeholders and stakes

In this step, the manager of the City-HUB model should identify his or her own stakes as well as (the stakes of) the relevant stakeholders, which need to be engaged in order to reach the identified policy goals. The endorsement of proposals to construct, expand or change an urban multimodal transport interchange by the stakeholders concerned will give decision makers and funding bodies confidence that the proposals can be implemented successfully. The real issue, however, is not about whether or not to involve these groups, but with what aims and by what means stakeholder involvement should take place. As a minimum, there is a need, at a technical level, to know about how travellers and other transport decision makers will respond to the various measures that might be included in a transport interchange.

- *Description of the good practice:* The validation case study of Utrecht showed that from the very start:
  - Stakes should be well defined in each step.
  - Relevant stakeholders should be committed to each step.
  - Roles and responsibilities for each of the involved stakeholders should be well defined and should be clear among stakeholders and within each stakeholder organisation.
  Pilot cases which considered interchanges as successful were focusing on multiple stakes and had clearly defined roles and responsibilities. Examples are the two railway stations of Lille Europe and Lille Flandres: a partnership with all the actors involved in the interchange, which exists, in concert between stakeholders, for long-term (10 years) urban planning in this 'railway stations triangle'. Euralille SPL have meetings with LMCU, the town of Lille, RFF (rail track owner), SNCF (train operator), Gare & Connexions (a subsidiary company of SNCF for the stations' organisation), the Nord Pas de Calais Region (TER transport authority), Transpole (Urban Public Transport operator that belongs to KEOLIS, a subsidiary company of SNCF), Unibail (which owns the shopping mall) and the Département of North (for some roads that belong to the Département) to combine urban development and transport facilities.
- *For whom is it good practice?* Identifying stakes and stakeholders is vital for any user and any type of interchange.
- *For what kind of interchanges is the good practice suitable?* Identification of stakeholders and stakes is necessary for all types of

interchanges, but it is most important for more complex interchanges, in which multiple stakeholders should be involved.

8.5.1.4 Define functions and logistics of the interchange

The main purpose for any interchange is to provide efficient and seamless transfers between modes. This means that the distances between modes should be kept to a minimum and provide good links for transfers. It also means that in the planning stage, the responsible stakeholders should consider current and future developments in terms of passenger demand, service demand and general development connected to the public transport system. These aspects have a major influence on the overall design of the interchange and set the basic requirements for the size and number of platforms necessary. Key questions are, for instance, whether and how many of the buses should end/start at the interchange and what the distances between modes should be. Transfers should be as seamless as possible in order to reduce the interchange penalty. Interchanges located in central areas can also potentially occupy large areas. Therefore, it is of vital importance that functions and logistics at the interchange are well planned. Central areas can have great potential for mixed use and high potential for alternative use. A second issue is to plan the overall logistics at the interchange and the interaction with services well.

The validation case studies show examples of interchanges that have long distances between modes and situations where the different parts of the interchange have been planned ad hoc. Moreover, the interchange's role in the transport system is also not well planned.

- *Description of the good practice:* The definition of the functions of interchanges should be clearly laid out at the outset. The main functions to consider are transport (future size, frequencies, passengers and link to the overall transport network), logistics and retailing.

  When constructing and refurbishing the Moncloa interchange in Madrid, the responsible stakeholders took great care in planning how the interchange should correspond with overall developments in the transport system. A common challenge for large cities is that there is reduced accessibility (for buses) to city centres due to heavy congestion. In order to address such a challenge, the authorities constructed six interchanges in the vicinity of the city centre. The primary goal is to offer transfer from buses to the urban network. Travellers therefore have to transfer to the urban network in order to decrease travel time and reduce congestion in central areas. Consequently, they have developed a coherent transport system in which the interchanges play a key role.
- *For whom is it a good practice?* Travellers will be the main beneficiary of a well-defined interchange. This could reduce the distances between

modes and provide seamless travel that reduces travel time and stress connected to transferring between modes. A well-functioning interchange is also important for operators and owners of the interchange since it can increase the use and popularity of public transport. A carefully planned transport system can also reduce the costs of transport services allowing an increase in journey frequencies.

- *For what kind of interchanges is the good practice suitable?* It is well documented that travellers would prefer to travel directly without needing to transfer between modes when using public transport. Consequently, it is of vital importance to reduce the interchange penalty. All types of interchanges should therefore take great care in defining the functions and logistics at the interchange. The size and organisation of functions are highly dependent on the local context.

8.5.1.5 Ensure energy efficiency

Reducing emissions from cars and buses has traditionally been the environmental focus of public interchanges. Less emphasis has been placed on the energy efficiency of the construction of interchanges. Consequently, the case studies revealed that the majority of the interchanges did not employ a low and zero carbon energy construction criteria. There are also examples of interchanges that have not implemented any significant strategies to increase energy efficiency in the operation of the interchange.

However, sustainable interchange design and energy efficiency are topics which receive increased attention from policymakers and authorities. These issues are not only relevant for the environment, but can also have positive impacts on social and economic elements. The responsible actors should take into account that interchanges consume a large amount of energy over their lifetime. A low carbon energy strategy and a focus on energy efficiency are needed for interchanges to reduce their environmental impact both locally and globally. In addition, it can reduce operating costs for the managers of the building.

- *Description of the good practice:* Key environmental impacts of an interchange are related to building construction and the energy efficiency that could be achieved by its construction and design. Buildings and other infrastructure can be constructed with energy efficiency as one of the primary considerations. This includes reducing the use of energy through, for example, using daylight and high efficiency energy systems, as well as the inclusion of renewable energy technologies. This is especially important in the planning of new interchanges. In Oslo, there are plans to move the current bus terminal above the rail tracks. According to the interviews, the responsible stakeholder aims to classify the new interchange with 'BREEAM-NOR excellent' status. BREEAM (Building Research Establishment's Environmental

Assessment Method) classifies buildings according to the following levels: pass, good, very good, excellent and outstanding. In order to receive a BREEAM excellent result, they need to document that the various criteria have been fulfilled. If the interchange receives a BREEAM excellent result, it would become one of Norway's most environmentally friendly buildings.

However, there are also examples of existing interchanges that have implemented strategies to improve the environmental impact. Kamppi uses the exhaust air from the passenger areas to warm up the bus platform which does not have any heating system. Air filters have been added to the roof of the terminal area (which is an indoor space) to improve air quality. Kamppi and Oslo bus terminal Vaterland also have doors to the bus platforms which only open while buses are departing. This can be a positive measure for security (avoiding passengers encroaching on areas where buses are moving), but also for energy efficiency since it reduces the unnecessary leakage of heat. Kamppi also has a system for monitoring the air quality and regulates the idling of buses.

Moreover, for the rail stations in Lille, there is a target of 20% energy reduction by 2022 compared to 2012 by the use of teleprocessing of the lighting.

- *For whom is it a good practice?* Ensuring energy efficiency and reducing the carbon footprint are first and foremost beneficial for the interchange owners. The responsible stakeholders can have lower operating costs by using less energy. By having high standards for energy efficiency and the use of carbon energy, they can also market environmentally friendly behaviour and show that they take corporate social responsibility seriously.
- *For what kind of interchanges is the good practice suitable?* For the planning of new interchanges, environmental considerations in construction and operation should be a primary concern. It is difficult to obtain the same energy efficiency in existing interchanges as in new interchanges, and large investments may be required to obtain low energy use and low carbon emissions. Nevertheless, existing interchanges can implement strategies that reduce the use of energy.

8.5.2 Validation and deployment: managing an interchange

8.5.2.1 Map out opportunities to link up with the private sector/provide a financial plan and welfare evaluation

Proper mapping of private sector interests (identification of stakes and stakeholders) in the interchange in various forms (investors, service providers, subcontractors, etc.) can effectively scale up resources available for interchange management from the time of project preparation. On the other hand, cooperation between stakeholders means that besides financial consequences, consequences for common decision-making also arise. Therefore, at the end of this step, the authority, as well as the involved stakeholders, should have a decision or an agreement on the chosen business model, including the governance structure.

- *Description of the good practice:* In Spain, the city of Madrid has promoted the use of public transportation by the adoption of different measures. The construction of intermodal exchange stations has been one of the most prominent investments in improving the physical connection between metropolitan bus services and the subway system. Di Ciommo et al. (2009) evaluated the effects of such an interchange in Madrid on the affected stakeholders: travellers, operators, interchange managers, the government and the urban environment (people living near the interchange). The authors applied a welfare perspective to the Avenida de America Interchange in the city of Madrid. The financial analysis shows that public subsidies are not needed in order to have interchange construction or redevelopment realised.

  The analysis of the case study of the Avenida de America compares costs with benefits. The main benefits result from the travel time gains for metropolitan and urban buses (between 3 and 7.5 min/trip). This leads to travel time savings for both travellers and bus operators. Although the authors conclude that this has resulted in an increase in bus demand, it is not clear whether the users have given up car use (which may lead to environmental benefits as well). Transportation fares did not change. The bus operators have to pay a fee for using the interchange to the owner of the interchange (IES concessionaire). The reduction in operating costs compensates for the fees. The benefits for the concessionaire are not only these fees, but also parking revenues and the commercial rents from shops. In sum, the implementation of this interchange has a positive welfare outcome. It is concluded that it is a win-win strategy with the interchange being financed by private capital only.
- *For whom is it good practice?* The viability check is good practice that benefits the stakeholders involved in the process. The output will also benefit travellers.
- *For what kind of interchanges is the good practice suitable?:* The viability check is vital for any user and any type of interchange.

8.5.2.2 Interchange management plan

An interchange management plan should be developed, and signed up to, which clearly states the roles and responsibilities of all those involved in the interchange through the different stages (i.e. design, planning, construction and operation). This is similar in approach to the Station Travel Plans, which is a partnership between rail operators, local authorities, bus operators and other stakeholders to improve access to stations.* This should avoid fragmented decision-making and importantly ensure that the priorities of all of the different stakeholders are considered. It should also set out the priorities for future development of the interchange. Interchanges are dynamic facilities and the management plan should reflect this; it should be updated on a regular basis.

- *Description of the good practice:* The governance model should clearly define who bears liability and financial responsibility for certain operations and what should be done in the case of significant revisions to the existing operational model. These could be caused by regulations on accessibility, emissions, ventilation requirements and so on. Collaborative efforts of all stakeholders would be also needed for serious upgrades and refurbishments. Again, a detailed understanding of the financial model is required in determining each stakeholder's role in these processes.

  In the case of Madrid, the interchange management plan was developed, and signed up to, which clearly states the roles and responsibilities of all those involved in the interchange through the different stages.
- *For whom is it good practice?* An interchange management plan is vital for any practitioners who will use the City-HUB model. It makes roles and responsibilities clear.
- *For what kind of interchanges is the good practice suitable?* An interchange management plan is vital for any type of interchange but it is most important for more complex interchanges in which multiple stakeholders are involved.

8.5.2.3 Fix complementary policies and development plan

The success of an interchange is not only dependent on the standards within the area of responsibility for the interchange manager; it also depends on complementary policies for a wide range of aspects.

- *Description of the good practice:* In Moncloa, the interchange development constituted a vital part of an overall plan for larger refurbishment or development. In Lille, there is a development plan for city integration, as well as a steering committee with elected members and

---

* See 'Guidance on the implementation of Station Travel Plans' available from www.station-travelplans.com.

technical committees to develop transport, urban and development policies.

This illustrates the necessity of political willingness to implement complementary policies in order to provide a better environment and to boost the transport investment and obtain economic development. These complementary policies can facilitate implementing interchanges as part of an overall larger integrated policy and/or plan aiming at (re)developing linked economic activities and urban function (re)development. Moreover, 'policy design also has a crucial role in influencing and strengthening the potential impact of transport infrastructure investment on local economic development' (Banister and Berechman 2001).

- *For whom is it a good practice?* Securing complementary policies and having a development plan for land use can be beneficial for all involved stakeholders at the interchange. Travellers can experience better accessibility and way-finding, thus reducing travel time and making the journey less stressful. Operators should ideally not experience congestion that causes delays. A carefully planned policy should take such considerations into account. This would again increase the attractiveness of public transport. Public authorities and business can also benefit from such policies. Land use and development plans can increase the attractiveness in the area and consequently attract businesses and offices to establish themselves in adjacent areas. The interchange owners can thus experience increased travellers and income.

- *For what kind of interchanges is the good practice suitable?* These aspects are important for all types of interchanges. However, the possible positive effects are largest in cases where the interchange has many passengers and is located in areas where development can take place.

8.5.3 Monitoring and assessment for users

The case studies showed that the majority of interchanges are not regularly conducting surveys in order to receive feedback. The same applies when it comes to having a system that assesses performance efficiency and effectiveness. Consequently, the interchange managers have not institutionalised a method for receiving feedback regarding user experiences at the interchange. Interchange managers can also set up a network consisting of operators and users for discussing issues of relevance to the operation of the interchange. The case studies showed that not all interchanges have such a system in place.

At the same time, the case studies illustrated that, in general, there is significant room for improvements at the interchanges. Without receiving feedback or monitoring performance, the interchange managers do not necessarily acquire important information on what is necessary to improve or the users' feedback about using the interchange.

- *Description of the good practice:* It is necessary to receive feedback from users and evaluate performance at the interchange. Feedback can be institutionalised by having regular meetings with operators and/or travellers at the interchange. It can also be done by conducting regular surveys among the travellers and implementing a system that can assess performance efficiency and effectiveness.

  By doing so, the interchange can get constant feedback on areas that function well or badly. The responsible stakeholders then have the possibility to evaluate how the interchange is functioning and implement possible solutions to improve the status. Ultimately, whether an interchange is regarded as successful or not is to a large extent evaluated by user satisfaction. Gathering information about the users' perception of various elements can therefore be an important means of ensuring that any redevelopments satisfy their interests and needs. User surveys can thus highlight elements that have not been adequately taken into account. Conducting several studies of various interchanges makes it possible to develop a better understanding of how various solutions are regarded. An indirect effect of undertaking these types of surveys might be a better understanding and capacity building for those working with interchanges.

  The Moncloa interchange in Madrid conducted a User Perceived Quality and Satisfaction Survey prior to starting the last refurbishment of the interchange (in 2008). The purpose was to develop a quality assurance plan and a set of actions for the interchange. Two topics were included in the user questionnaire: aspects which were regarded as important at an interchange, and users' degree of satisfaction with services. An analysis of the results showed that security, functionality of services, information and the station's general appearance were regarded as the most relevant aspects.

  In Lille, regular surveys on the level of satisfaction among the clients are also conducted. In Helsinki, regular common surveys concerning the level of service for the whole regional and local public transport are conducted. The surveys are not directly linked to the interchanges, but some special studies have also been made about interchange safety and security issues.

- *For whom is it a good practice?* Good practice is first and foremost aimed at interchange managers. By monitoring performance and receiving feedback from users, they have the possibility to assess the current operation and use this as a point of departure for implementing improvements at the interchange. The output can be improved facilities and cooperation among the users.

- *For what kind of interchanges is the good practice suitable?* Monitoring performance and receiving feedback from users is relevant for all types of interchanges. It is especially important for larger interchanges with many users and stakeholders involved. This is because the operation becomes more complex. In addition, the importance of having a well-functioning interchange increases with the number of passengers that need to be accommodated.

8.5.4 Interchange design and accessibility

The key goal for any interchange is to provide travellers with good accessibility and easy transfers. The design of the interchange significantly influences how long it takes to transfer between modes and how easy it is to orient within the interchange. The design also has important consequences for the pedestrian flow and level of conflict between travellers. All these factors influence the attractiveness of an interchange.

The case studies have revealed that there are challenges connected to, for example:

- Gate lines could easily become overcrowded during rush hour.
- Conflicts between those queuing for tickets and those queuing at the barriers.
- A fairly small area for ticket office and ticket machines.
- Narrow connections.

This means that travellers experience several barriers that increase their travel time and lead to uncomfortable public transport travel.

- *Description of the good practice:* An instrument that can amend such problems is an analysis of the pedestrian environment surrounding the interchange, including assessing the level of comfort. Such an approach has been undertaken in Ilford in preparation for the station redevelopment and urban realm improvements as part of the Crossrail development. The analysis concluded that the immediate entrances to the station were comfortable, but other areas were rated 'unacceptable' due to narrow footways and crossings, advertising signs and telephone boxes. Forecasts show that increased passenger flows are expected as a result of the Crossrail development which will result in lower levels of accessibility.
- *For whom is it a good practice?* By carrying out this analysis, the responsible stakeholders have the necessary information for implementing measures to enable users to find their way in the shortest time period possible by avoiding any barriers which increase travel time. This is an example of the benefits of using formal pedestrian

audit methods when planning a major redevelopment to identify and prioritise where improvements are needed. This will help to ensure that walking accessibility is improved once the station is redeveloped. Transport for London has a toolkit of pedestrian assessment guidance, applied alongside its guidance on the design of transport interchanges. In the end, the travellers will benefit from having an improved interchange that better facilitates transfers between modes.

- *For what kind of interchanges is the good practice suitable?* All types of interchanges can benefit from analysing pedestrian flow and accessibility. However, for larger interchanges, it becomes of greater importance to ensure the design supports efficient transfers.

8.5.5 Facilities and services

8.5.5.1 Facilities and retail services

Shopping facilities at the interchange provides travellers the opportunity to use their waiting time more productively. This can be an important element in reducing the interchange penalty and might be particularly important for travellers with rather long waiting times. However, it can also be important to take into account that passengers have to conduct several errands each day. Having services located at interchanges provides passengers with the opportunity to shop for necessary commodities during their regular travel and can thereby be an incentive for increased use of public transport.

- *Description of the good practice:* Kamppi interchange is centrally located in downtown Helsinki. The area of the interchange/shopping centre is approximately 4 hectares. In addition to the shopping centre, there are also offices and flats in the same building complex. A total of 170 businesses operate in the Kamppi shopping centre (May 2013). They have a wide selection of businesses, including 106 stores, 35 restaurants and cafés and 29 services such as beauty salons, gyms, banks and laundry. The gross lettable area in the Kamppi shopping centre is 43,000 m^2.

  The interchange station of Moncloa includes various shops and other commercial services. There are some restaurants, cafés, snacks and candy shops, general stores, newspapers kiosks, book shops and a tobacco shop. Often, the interchange is used like an advertising window for promoting new technological products (i.e. mobiles). There are also various vending machines that offer products like flowers or medicines. There are some cash machines.

  The quality of shops, restaurants and services could vary from a typical local restaurant to fast food.

- *For whom is it a good practice?* Facilities at an interchange should be attractive, making the space an enjoyable place to be and reducing the interchange penalty. Value of time generally decreases when travellers are waiting in the interchange and only have low quality or few facilities. This is crucial for travellers who could expect to spend some time waiting at the interchange. For such travellers it is important to provide enjoyable areas to relax or make better use of their time.
- *For what kind of interchanges is the good practice suitable?* The need to offer facilities and services varies between interchanges. The interchange needs to be of a sufficient size to offer a wide range of shopping facilities. However, all kinds of interchanges should aim at making waiting time pleasant for travellers.

8.5.5.2 Quality of services

The value of time on journeys can often be improved by providing travellers with the opportunity to use their time valuably, such as shopping or eating. It is also important to offer waiting rooms which give travellers protection against weather and noise. This is especially necessary during cold winter months. At a minimum, there should be protection against snow and rain. Another important feature, which is becoming more important, is the provision of internet through Wi-Fi. In recent years, there has been rapid technological development which offers travellers new opportunities for browsing the internet, checking emails, and so on by using laptops, tablets and mobile phones. Being able to work or read news on these devices is a factor that could help to offset the advantage that might be seen of using a car.

The quality of services and compliance with regulations can be supervised and monitored by interchange employees. This could help to ensure high standards of retailer quality, as well as avoiding challenges connected to, for example, the fact that facilities are not located in logical progression. Feedback from passengers and customers can also be an important source of information in order to provide better services. Customers can make suggestions for improvements and state their satisfaction with various aspects at the interchange. Feedback from passengers and customers can also provide better capacity.

- *Description of the good practice:* Moncloa Interchange is mainly located underground. Consequently, its users are protected against bad weather conditions. The passengers waiting for the bus are separated from the bus bays by screen enclosures; they are waiting in an open space. It is a quiet space because of the separation from the noise of bus traffic. Finally, the open spaces of the platforms are kept at a pleasurable temperature with regularly refreshed air.

There is Wi–Fi access in the commercial areas and on each platform for buses.

In Moncloa, the concessionaire authorises which services and activities are allowed to take place within the interchange. It is up to the concessionaire to set the standards for the services provided and thus they have control over deciding what kind of services are allowed (securing proper distribution of types of businesses) and the standards provided, for example, opening times, comfort and distribution of merchandise.

Kamppi interchange is an indoors interchange as well. The buses are separated from the waiting area by glass walls and doors that only open when the buses are about to depart.

The railway stations of Lille Europe and Lille Flandres both have two types of waiting rooms:

- The 'Salon Grand Voyageurs'.
- A 'Waiting Room' for all passengers.

The Salon Grand Voyageurs is a special waiting room at the disposal of passengers who have the SNCF Fidelity Card. In Lille Europe, this room provides facilities like a ticket desk with a hostess, screens with information on departures and arrivals, eight comfortable arm-chairs and ten plastic chairs, a table with plugs and free hot drinks, Wi-Fi, press magazines and access to a WC. In Lille Flandres, the Salon Grand Voyageurs provides almost the same facilities, but there are more seats and a bigger table with plugs for battery charging. Waiting rooms for all passengers are well lit and equipped in both stations with wooden chairs, departure/arrival times and little tables with plugs, but there is no air conditioning.

There is unlimited free Wi-Fi available in the Gare Lille Flandres interchange, and it is also possible to access the internet using a credit card on dedicated machines located in both stations.

- *For whom is it a good practice?* The image of an interchange is affected by the quality of service provided within it and also in the surrounding area. High standards connected to the design of the interchange and its facilities are likely to have a positive impact on travellers' satisfaction and their value of time. It is particularly important to reduce the perceived time spent on a journey.

  All passengers can benefit from these services if they have the required quality. Intermodal services also have indirect impacts; for example, providing free Wi-Fi can facilitate fewer queues at ticket machines if it is possible to buy tickets online through mobile phones and other devices.
- *For what kind of interchanges is the good practice suitable?* It is the interest of any kind of interchange to provide intermodal services of the required quality. Among other things, this can make the interchanges more attractive.

8.5.6 Information

Intelligent information systems: Information is a necessity for public transport users and important to consider when promoting the increased use of an interchange. Lack of information increases barriers for interchange users. This is especially problematic for less frequent public transport users. Seamless intermodal journeys require integrated information across modes since this promotes both time and effort savings (Grotenhuis et al. 2007). Such practices are especially important if the interchange also lacks direct customer services, such as an information desk.

In the last decade, the rapid technological development of real-time information has been observed. The pilot case studies revealed that some regions have developed advanced systems, while others still have some way to go before they can offer such services. For instance, at Moncloa, the various displays only provide information related to a particular mode of transport and are not integrated across modes.

- *Description of the good practice:* At Ilford Railway Station in London, the level of information provision varies between the different parts of the interchange. Within the train station, the level of information is good, with both timetables and real-time information and other passenger information being provided on the platform and within the front (main) entrance. Passengers are also able to get real-time information online through the National Rail website or mobile phone app. Within the main entrance hall, screens show details for each of the platforms, with an additional screen showing various items of passenger information, including safety messages, information on the service status of the London Underground and information on the service status of Greater Anglia's services (Figure 8.1). Public announcements are made on the platforms.

  Various websites, including National Rail and Greater Anglia, provide information that allows users to plan their rail journey. Transport for London's Journey Planner allows users to plan their journeys in advance – combining the various modes available in London and also allowing those with mobility issues to select whether they need 'step-free access'.

  At the bus stops, the level of information is also good, with both timetables and real-time information being provided, along with maps of local bus services. Passengers are also able to access real-time information online through Transport for London's website. Some information on rail services is also provided on the outside of the station building, as are a number of maps of the local area.

- *For whom is it a good practice?* Prior planning is often crucial for users of public transport and a key factor for promoting increased use of public transport. It is easier to plan and optimise intermodal trips

Figure 8.1 Passenger information screen at Ilford Railway Station, London (UK). (Courtesy of Jan Spousta.)

if journey planners provide information about journeys across modes. Prior knowledge about journeys for all modes can make passengers less stressed and facilitate better use of their spare time (this links to the importance of the services provided within interchanges).

• *For what kind of interchanges is the good practice suitable?* Beyond online journey planner applications, it is essential that interchanges provide sufficient travel information for passengers across modes. Therefore, each kind of interchange can benefit from these good practices and should use such systems.

8.5.7 Safety and security

The passenger surveys (see Appendix III) documented that safety and security are regarded as key factors for passengers at interchanges. Safety and security in transport are also among the primary concerns of the European Union. Safety encompasses areas such as minimising the risks for collisions, conflicts and accidents for operators and passengers at interchanges. It also involves compliance with relevant safety standards. Security in the transport sector refers to serious crimes, for instance, terrorist attacks (less frequent) and other crimes, such as robberies and theft (more frequent).

Coccia et al. (1999) identified four main types of fear expressed by users in terms of security:

• Physical attack and sexual assault.
• Theft of cars, car parts and car radios.

- Theft of bicycles and bicycle parts.
- Vandalism to vehicles and buildings.

The case studies show several examples connected to unsafe design that increase the risk of accidents. For instance, in Köbánya-Kispest, there are several pedestrian crossings across roads used by buses. Ilford also has secluded areas at the interchange that make travellers feel less secure in the hours of darkness, especially women and vulnerable people.

- *Description of the good practice:* Interchanges can implement a range of assets that can reduce the risks from accidents and man-made threats. Crime prevention, the design of usable and secure facilities and the use of various means to monitor the interchange are some examples. There are also (usually) numerous legal requirements that the stakeholders need to comply with. The overall safety and security standards at an interchange are thus dependent on a combination of measures.

  The case studies have shown several good practices connected to these issues. First, one key aspect is reducing the risk of accidents by keeping the flow of vehicles and passengers separate. This can be done through, for example, bus bays that ensure that passengers do not use areas where the buses operate (Moncloa) or having doors that only open when the buses are about to depart (Kamppi and Vaterland).

  Second, great care should be taken into considering safety and security issues in the design. The Moncloa interchange conducted a specific study of air extraction and ventilation in case of fire. Prior to the construction of the interchange, simulations of passenger movements during evacuations were also conducted. These studies influenced the design, location and number of emergency exits, achieving building evacuations in less than six minutes during peak hours.

  Third, emergency management is vital in order to achieve a safe and efficient handling of an emergency. In Moncloa, the interchange has exclusive facilities for the police. They also have a fireman responsible for fire safety and evacuations. The interchange will therefore always have highly competent personnel in cases of emergency or for issues involving safety and security.

  Fourth, monitoring, staff available and secure parking for cars and bicycles are necessary to reduce the (fear of) theft of cars and bicycles. There are several examples of interchanges that use CCTV extensively and have security personnel to combat such problems.
- For whom is it a good practice? Safety and security is a vital issue for passengers. The case studies have identified several aspects that can make users of the interchange feel less secure. Some of the issues involve secluded areas that can make passengers feel unsafe in the

hours of darkness, especially women and vulnerable people. Feeling insecure can lead to lower levels of comfort and make public transport less attractive.

Safety and security is also important for operators. At Vaterland, there have been newspaper articles about bus drivers threatening to stop using the interchange. They were especially sceptical about the lack of safeguarding for drivers, employees and passengers. According to the safety deputy for one of the operators, buses suffer substantial damage yearly and bus drivers fear hitting people when they are backing out of the docking platform.

- *For what kind of interchanges is the good practice suitable?* Both small and large interchanges need to comply with the current safety and regulation standards. However, the level of safety and security is probably dependent on the type of interchange. Larger interchanges have more passengers and much higher frequencies of buses/trains. The risks for accidents can be higher compared to smaller interchanges and the travellers can have more fears about crime and overcrowding. All types of interchanges should however reduce the possibility for accidents, vandalism and theft.

  The possibilities for terrorist attacks are significantly larger at interchanges that have a larger number of passengers. This calls for taking into account sufficient safety and security measures in the design and operation of such types of interchanges.

8.5.8 Coordination of intermodal transport

8.5.8.1 *Intermodal coordination*

Coordination between modes is a key aspect for any interchange. Public transport attractiveness is closely linked to the relative travel time compared to that of cars, making transfer time an important element of journeys. The responsibility for operating public transport might be divided across a number of stakeholders, thereby making it more difficult to strive for better integration of routes. The responsibility for achieving good coordination usually lies within the operators' responsibility. The interchange should mainly facilitate transfers.

The need to coordinate intermodal transport depends on the service frequency for a particular line. High frequencies reduce the need for cooperation between operators and the need for coordination. This is partly because high frequencies lead to a network effect in which travellers forget the timetable. Preplanning and coordination become of greater importance if there is a long headway between modes of transport.

The pilot cases studied the average waiting times across modes. The results showed great variations between the five interchanges. For some transfers, the average waiting time was estimated to be more than 20 min.

This means that intermodal public transport becomes less attractive since waiting time can constitute a significant amount of the total travel time.

- *Description of the good practice:* The Moncloa interchange achieves excellent results in terms of increased demand and improved journey times for both users and the transport companies.

  The interchange, situated at the northern edge of Madrid, but in a built-up area, provides a gateway to the city for over 287,000 people per day. Buses and metro trains depart every 5–10 min during the peak hour. Thus, travellers have, in general, short transfer times. The need for coordination also becomes of less importance. The same applies for the case study from Kamppi. There are at least five departures every hour, making it easy to transfer. The same is true for transfers between metro/train and bus/metro at Kőbánya-Kispest.

- *For whom is it a good practice?* Coordination between modes will mainly benefit travellers who use public transport. High frequencies lead to a network effect in which travellers forget the timetable. Preplanning also becomes less important. This aspect can ultimately influence the operators. Better public transport services can make public transport more attractive and reduce the possible comparative advantage of cars.

- *For what kind of interchanges is the good practice suitable?* This aspect should be key for any kind of interchange, regardless of size or location.

8.5.8.2 Integrated ticketing

Electronic ticketing across modes is essential. Electronic ticketing makes it possible to have one ticket that can be used on various modes within a region. Electronic ticketing can thus reduce barriers connected to the problems of buying tickets, as well as saving time, for example, when boarding public transport, and as a result make public transport more attractive.

Integrated ticketing is not actually a question of interchange points, but a key issue of the use of public transportation and acceptance of intermodality, that is, changes during a trip. Integrated ticketing thus leads to an overall increase in the use of interchange points, which makes the services and quality of the terminals and other interchange points extremely important.

Ideally, the administrative boundaries between operators should not be visible to passengers, although achieving this will require discussions and negotiations beyond the scope of the interchange facility zone in question. Benefits to passengers in terms of integrated ticketing can be achieved, for example, through ensuring that ticket offices and ticket machines sell tickets for all services using that interchange facility or zone, rather than having different facilities for each operator. Staff should be trained so that they accept all tickets, not just those provided by their company. It is also important that different fare areas are clearly defined and that passengers know when they are moving from one operating environment to another. This could be achieved by using gate lines, positioning of smart ticket readers and the use of advisory signing.

- *Description of the good practice:* In Moncloa, as of May 2012, the new smart Public Transport Travel Card has been put in use. The implementation will be undertaken progressively, starting with the travel card in Zone A.

  Based on Radio-Frequency Identification (RFID) technology, these contactless cards offer numerous benefits compared to conventional contact-based tickets. As the ticket validation is carried out without direct contact with a reader, public transportation organisations are able to realise cost savings through reduced maintenance efforts for heavily stressed mechanical equipment in conventional, contact-based mass transportation systems.

  Furthermore, the ticket validation process is much quicker than manual stamping, thus reducing waiting queues and offering a convenient entry process for passengers. RFID-based public transportation applications help to significantly increase efficiency for several million passengers per day that need to be equipped with reliable and convenient access solutions (Figure 8.2).

  At Ilford, the integrated ticketing is successful as passengers are able to use a smart ticket (the Oyster Card) for both rail and bus journeys made from the interchange. This could nevertheless be further improved if passengers were also able to pay for car parking and taxis – the other

(a)   (b)

Figure 8.2 Ticket validation: King's Cross, London, United Kingdom (a) and Lille Flandres (b). (Courtesy of Andres Monzon (a) and Jan Vasicek (b).)

modes currently available from this interchange. Ticket purchasing for this interchange varies between the modes. Within the train station, tickets can be purchased from the ticket counter and ticket machines in the main entrance hall. Tickets are also available to buy online prior to travel – they can be collected from the ticket machines, delivered by post or printed by the passenger. For rail journeys within London (and out as far as Shenfield), it is also possible for passengers to use an Oyster Card. For the bus, tickets can be purchased from local newsagents or on the bus. It is also possible for passengers to pay using an Oyster Card or by using a contactless payment card.

In the Kamppi interchange, travel cards are used for Helsinki local transport. Paying for the trip with the travel card is cheaper than using cash. Mobile tickets are available for the metro and some bus lines connecting to the metro. The card offers easy payment and validation, which speeds up the process and saves time for the traveller. It is also more convenient because there is no need to carry change for the ticket.

- *For whom is it a good practice?* Integrated ticketing is good for both providers and travellers. Speaking of the providers, the integrated ticketing allows them to contribute, helps to share the revenue and can save infrastructure costs, since there is no need to implement different validating infrastructures (gates, scanners, etc.) and the maintenance costs are lower. As a result, the more integrated the tariff system, the more people will use it.

  The other group of affected stakeholders are the people who are using the infrastructure. An undoubted advantage of the integrated ticketing is saving time for the traveller, in many ways. There is no need to worry about how many different tickets are needed for a trip; one ticket will be used on the whole trip, which makes the process convenient. Also, the ticket could be purchased via the internet or even a credit card with PayPass technology, and it could be used as a ticket or pass. This feature saves time for the passenger. Using contactless cards speeds up the process, which means that more people could gain access to the service.

- *For what kind of interchanges is the good practice suitable?* Integrated ticketing is a fairly independent feature in the aspect of intermodality. There are several circumstances which could affect integrated ticketing, and this system is suitable for every type of intermodal terminals. However, the operation of an integrated ticketing system depends not on the interchange, but on the need for contributions from the local government and the service providers, among others. Thus, the implementation or improvement of integrated ticketing is far beyond the competence of any interchange, but inevitably makes travelling more convenient, in all situations, for all interchanges.

8.6 CITY-HUB MODEL AND ITS TRANSFERABILITY

The City-HUB project has utilised two sets of case studies as part of the work. At an early stage of the project, a set of five pilot case studies was used to assess good and bad practices and improvement potential in urban interchanges. Interviews with practitioners of 21 additional European urban interchanges have largely contributed to define the City-HUB model, the related checklists and the good practice. The lessons learnt from these analysed case studies served as input validation case study analysis. Six validation case studies were selected and used to test the model for specific interchanges across Europe.

The model was supported by a set of checklists covering a broad spectrum of interchange qualities, including different categories varying from management and ownership issues to key interchange properties that are important for travellers.

For all validation case studies, the interviewees have expressed an interest in the City-HUB model and have stressed the need for such a model. From the Paseo de Gracia case study in Barcelona, it was reported that it is very useful to have such a model because not enough guides for designing or assessing interchanges exist. The variety of cases studies and types of actors consulted reveals, however, that there may be different needs, and all of these cannot be satisfied with one model.

On the one hand, it was requested that the model be made more specific, while others have emphasised that it is good to have a common, more holistic model for interchanges. There may also be a need to adapt the model to different local circumstances, and varying emphasis may be placed on different parts of the model depending on the stakeholders involved and what situation the model is supposed to contribute to. The City-HUB model should thus be understood as a framework for analysing different interchanges across Europe and as a guide to good processes.

During the development of the City-HUB model, it was analysed whether it would be possible to establish two different versions of the model; one for processes where new interchanges are constructed and one for existing interchanges where improvements or upgrades are being planned. The conclusion was that it would *not* be wise to make two separate models, one reason being that sometimes a clear line cannot be drawn between upgraded and new facilities. The checklist that was used to assess interchange properties is particularly appropriate for the evaluation of existing interchanges to identify room for improvement. For new interchanges, the checklist is rather an indication of which elements should be considered in the planning (see Miskolc interchange validation case study in Hungary).

Feedback from the case study interviewees also related to the discussion between different stakeholders and the need to define clear roles and responsibilities. The degree of stakeholder complexity and the roles allocated to each stakeholder varies between countries and regions, between owner-ship structures and between interchange types and modal combinations. Therefore, the City-HUB model does not deal with exactly which role each stakeholder should have, but the key message is that the roles have to be clearly defined. The consequence is that the improved City-HUB model takes into account the importance of having all roles and responsibilities clear from the very start.

The experiences from the use of the checklists were, in general, positive. From several interviewees, it was reported that the checklists seem very useful and that they can help identify important characteristics that would not otherwise be detected.

Good practices are regrouped following the main issues of an interchange: governance; key elements of the interchange (validation and deployment); and how interchanges can be made attractive for users (monitoring and assessment).

This benchmarking study, through the good practices analysis, provides guidelines for checking the current status of the existent interchange and defining the key aspects for planning a new interchange. The good practices are related to the three sequential steps of the City-HUB model (i.e. identification, validation and deployment, monitoring and assessment), and the transversal step of governance as related to the stakeholders' participation in designing, managing and maintaining an urban interchange.

The main results of this benchmarking activity contribute to the defini-tion of main recommendations for making urban interchanges an efficient transport node and a place to change modes of transport, but also a place to spend intermodal waiting time.

Chapter 9

Urban interchanges for efficient and sustainable mobility

Andres Monzon and Floridea Di Ciommo

CONTENTS

9.1 THE ANSWER OF INTERCHANGES TO COMPLEX URBAN MOBILITY

The EU Green Paper on urban mobility (EC 2007) points out the three challenges of sustainable urban mobility: reducing congestion; improving the quality of public transport services to attract trips from private cars; and promoting soft modes such as walking and cycling. But how to increase public transport patronage when urban sprawl is leading to more complex travel patterns, reducing the attractiveness and performance of public transport (Hernandez et al. 2015)? Public transport trip distances and durations are increasing in urban and metropolitan areas (Banister 2011).

As a consequence, a significant percentage of urban trips are becoming multimodal or multistage, connecting different public transport services in a chain to complete the trip from origin to destination. Therefore, transfers between modes or lines are now a significant structural element of integrated urban transport systems (Navarrete and Ortuzar 2013). Some authors (Guo 2008) studied how transfers between the interurban/commuter rail and metro networks affect the modal split in several cities. Navarrete and Ortuzar (2013) analysed the penalty of transferring between buses and metro in Santiago. Their results clearly indicated that travellers perceive their trips as penalised by transfer time.

Therefore, the solution is not an easy one. Mobility patterns are becoming more multimodal but the transfer penalty is considered important.

How can the public transport network be made competitive with cars in this difficult context?

This book has been dedicated to helping solve this problem, in particular in metropolitan conurbations. Why not convert the need for transfers into a key element for having more attractive public transport networks? This means making special treatment of transfer nodes in order to reduce the transfer penalty and counterbalance it with benefits for travellers, which means efficient urban transport interchanges are needed for sustainable mobility. Therefore, improved interchanges could be the answer, or at least part of the solution, to making public transport systems more attractive with a higher quality in comparison to cars.

However, this is not an easy task, because the design of interchanges depends on so many factors, as shown in previous chapters. The City-HUB model aims to provide the basis for making interchanges attractive to help public transport be more competitive in the complex mobility patterns of advanced cities.

9.2 FROM A 'NON-PLACE' TO A 'KEY-PLACE'

The French anthropologist Augé (1995) said that multimodal passenger interchanges are examples of anthropological 'non-places' from the social point of view. They are not places where social relations could be developed, and lack a sense of history and identity. However, Bertolini (2006) found that they are, on the contrary, places with clear spatial dimensions, but they have complex uses where there is no dominant group of users. The target of this book has been to understand who the users of interchanges are and what their needs might be. According to Edwards (2011), a transport interchange is a more complex transport facility than a conventional station that allows travellers to transfer from one mode to another. It said that there are two main definitions of interchange: one is related to the infrastructure vision, which means that an interchange would be a place where several modes of public transport interconnect, and the second vision is based on users – a place where people transfer between two or more public transport modes. The two visions are complementary, but both together do not provide the completed vision of the interchange. It is something more than a building where there are several transport modes, and where many people transfer between them. The challenge is to convert a 'non-place' into a key element of the transport system which provides clear added value for travellers. If this target is achieved, the weakest point of the system – the need to transfer – could be converted into a strong point to make public transport more attractive and competitive (Figure 9.1).

As Edwards (2011) stated in his excellent and well-illustrated book, interchanges are more than spaces where there are buses and rails and people

Figure 9.1 Inside Moncloa Interchange, Madrid, Spain. (Courtesy of Fiamma Perez.)

transferring. In many countries they are 'places' rather than 'spaces' that restructure urban centres or remodel cities.

Chapter 3 showed how to put together all the elements and dimensions of an interchange place: users, transport modes, services and facilities. These elements define the needs for space and the characteristics which have to be developed, taking into account local constraints. Depending on the size and number of activities, the interchange will became a clear landmark in the city, or at least a point of reference in the urban geography. But the process not only has 'indoor' impacts for users of the interchange. It also produces local impacts on the surrounding area.

Therefore, we need an integrated vision to consider the interchange 'habitat', including all the activities that take place inside the interchange building and the relationships between those activities located around and the ones inside.

9.3 INTEGRATED PLACE FRAMED IN A DEVELOPMENT PLAN

Integration is a multidimensional objective that should be pursued at various levels. The first level is among different transport modes to assure a seamless quality transfer between them. The time associated with distances between transport services are important to travellers, especially for those with mobility impairments, as each transfer can accumulate substantial time within a door-to-door journey. It should therefore be a goal of an interchange to have the distances between transport services as short as possible. The distance between modes is important, as is good connectivity and way-finding for facilitating multimodal trips. It is necessary to consider the degree of 'integration' between the different transport modes and not only the existence of different transport modes that may coexist.

For instance, does a unique interchange space exist where all the transport modes could display their own departure times?

A good connection among transfer paths and waiting areas is crucial for passengers. Time dedicated to transfer and waiting is clearly penalised in comparison with travel time. Depending on the type of transport services, the location of waiting areas, near the gates in urban services or separated in long-distance ones, would reduce the negative perception of waiting. In some cases, this could even be reversed, if the place is comfortable and has good communication and reliable displays of information; free Wi-Fi could make the stay perceived as positive. The interchange manager should coordinate all these different elements for favouring multimodal journeys.

Therefore, interchanges need to be designed so that they provide logical and easy passenger movement. Overcrowded areas and long queues to pass through ticket barriers reduce traveller comfort and efficiency. A poor-quality travel experience is one of the key reasons given for not choosing to travel by public transport.

The degree of integration between modes is implicitly dealt with in terms of information, way-finding and distances between modes. Integrated ticketing is a related issue. The aim of the City-HUB model is to capture such elements and provide relevant stakeholders with guidance relating to the main elements of interchange development. For instance, in cases where the interchange does not offer information across modes, the interchange assessment would show that it needs to be corrected. This illustrates the importance and challenge connected with analysing the interchange status.

The second level of integration is between the transport activities and the services and facilities provided at the interchange. Some of them are directly related to serving passengers, such as ticket offices and luggage handling. Coordination between services and resources within the interchange, as stated in the offer side of the business model, is necessary. Every element of the supply should be designed and allocated in the interchange space to improve efficiency for passengers. Shops and cafés should be located separate from transfer flows but not too far from them and the waiting areas. Ticket booths should be near the departure platforms with luggage storage nearby.

The third type of integration has a managerial character. It refers to how to manage all the activities in the interchange in a coordinated way, resolving any conflicts between the different activities and priorities. This goal requires an integral vision of everything, looking for the best coordination of activities, services and business within the interchange space along the different times of the day/week. It also includes the location of access/exit gates which affect, either positively or negatively, activities located in the vicinity of those crowded points. Criteria for way-finding and information provision could make the difference in this regard. It is strongly recommended that an independent interchange manager is employed to coordinate all the stakeholders and activities (Figure 9.2).

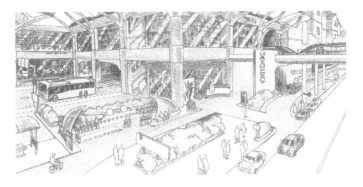

Figure 9.2 Integrated place. (Illustration by Divij Jhamb.)

Multimodal city-hubs have a clear influence on economic activities and land uses, particularly in the surrounding area. However, it is not clear how transport hubs establish their links with nearby economic and social activities. The size and role of the city-hubs (i.e. the interchange typology) within the transport network, providing direct links to the main cities, may explain the development around it. This wider dimension of the interchange requires a fourth level of integration: the interchange with the activities located around it. This kind of integration could be clearly fostered if a specific development plan exists which addresses all the interchange elements in a proper way.

The issue of integration between the interchange and the city, in particular connected to urban regeneration, has been included in the City-HUB model. It emphasises this aspect through the recommendation to 'identify policy goals/needs' at the beginning of the life cycle. An important point of the model is to evaluate the policy goals and plans that might affect the interchange in the final stage of the life cycle.

9.4 USERS' PERSPECTIVE RECOMMENDATIONS

Some elements of the interchange have a high importance for users generally, regardless of the context of the interchange, such as internal design and safety and security. Users' satisfaction regarding distances for transfer and facilities is generally high, although it influences to a lesser extent the global evaluation of the interchange, according to the answers to the pilot case study surveys. The availability of cash machines, telephone signals and Wi-Fi would significantly affect the overall evaluation of the interchange, whereas their absence would not.

The results obtained using the surveys carried out as the pilot case studies of the project show common key factors and attributes throughout the different surveyed interchanges. According to Hernandez and Monzon

(2016), these common key factors define an urban transport interchange under a dual approach: *as a transport node* and *as a place*. The interchange *as a transport node* is related to information provision – travel information and signposting – and transfer conditions – distances and coordination between operators. On the other hand, the interchange '*as a* place' is linked directly to aspects such as the design and image, indoor environmental quality, services and facilities and elements to improve the comfort of waiting time. In either case, safety and security aspects are essential for both approaches.

9.5 CITY-HUB MODEL FOR EXISTING OR NEW INTERCHANGES

The final outcome of the City-HUB project is to propose a model that includes parallel streams for the different stages of the City-HUB life cycle and the governance actions. This has been described in Chapter 7 and validated by a number of case studies.

The recommendation from the City-HUB model is to involve all stakeholders in the earliest stages of a project. Consideration should be given to the needs of different stakeholders, especially interchange users, with potential conflicts discussed and mitigated early in the planning or development process.

The functionalities of the interchange should be clearly identified in the light of the different needs and requirements. The result will be the selection of the most appropriate type of interchange and its characteristics. The building design and stakeholders' involvement represent key features of the interchange. The business model selection depends on the physical aspects of the interchange and its size, and the activities associated with the interchange, both inside and outside. Depending on its intrinsic characteristics and the economic activities generated, each interchange will therefore need a specific business model that could range from less integrated (cold/hot) to a fully integrated interchange.

Finally, a periodical assessment of the interchange is necessary to check the achievements of the policy goals in continuous dialogue with stakeholders. The City-HUB model includes a series of checklists to enable this to be undertaken comprehensively and consistently. Interchanges are not static entities, and therefore need to enhance their performance by updating their business model with innovative actions and quality assurance plans.

References

Abreu e Silva, J. and H. Bazrafshan. 2013. User satisfaction of intermodal transfer facilities in Lisbon, Portugal: Analysis with structural equations modeling. *Transportation Research Record: Journal of the Transportation Research Board* 2350(12): 102–110.

Adamos, G., E. Nathanail and E. Zacharaki. 2012. Developing a decision-making framework for collaborative practices in long-short distance transport interconnection. *Procedia—Social and Behavioral Sciences, Transport Research Arena 2012* 48: 2849–2858.

Analyse & Strategi. 2012. Organisering av eierskap og forvaltning av Vaterland Bussterminal AS.

Arnstein, S. R. 1969. A ladder of citizen participation. *Journal of the American Institute of Planners* 35(4): 216–224.

Augé, M. 1995. *Non-Places: Introduction to an Anthropology of Supermodernity.* London, England: Verso.

Auvinen, H., P. Christiansen and I. Markuceviciute. 2012. *Decision-Making Guidebook for the Interconnection between Short and Long-Distance Transport Networks.* Deliverable 6.3, CLOSER—Connecting Long and Short-distance networks for Efficient Transport. http://www.closer-project.eu/file/closer-d6-3/.

Avineri, E. 2012. On the use and potential of behavioural economics from the perspective of transport and climate change. *Journal of Transport Geography*, Special Section on Theoretical Perspectives on Climate Change Mitigation in Transport 24(9): 512–521.

Banister, D. 1999. Planning more to travel less: Land use and transport. *Town Planning Review* 70(3): 313.

Banister, D. 2011. The trilogy of distance, speed and time. *Journal of Transport Geography* 19(4): 950–959.

Banister, D. and Y. Berechman. 2001. Transport investment and the promotion of economic growth. *Journal of Transport Geography*, Mobility and Spatial Dynamics, 9(3): 209–218.

Bellman, R., C. E. Clark, D. G. Malcolm, C. J. Craft and F. M. Ricciardi. 1957. On the construction of a multi-stage, multi-person business game. *Operations Research* 5(4): 469–503.

Bertolini, L. 2006. Fostering urbanity in a mobile society: Linking concepts and practices. *Journal of Urban Design* 11(3): 319–334.

Breiman, L., J. Friedman, R. A. Olshen and C. J. Stone. 1984. *Classification and Regression Trees*. New York: Chapman and Hall/CRC.

Brons, M., M. Givoni and P. Rietveld. 2009. Access to railway stations and its potential in increasing rail use. *Transportation Research Part A: Policy and Practice* 43(2): 136–149.

Cervero, R. and M. Duncan. 2002. Transit's value-added effect: Light and commuter rail services and commercial land values. *Transportation Research Record* 1805: 8–15.

Cervero, R. and J. Murakami. 2009. Rail and property development in Hong Kong: Experiences and extensions. *Urban Studies* 46(10): 2019–2043.

Cervero, R., Parsons Brinckerhoff Quade & Douglas, Inc., Howard/Stein-Hudson Associates, Inc., and J. Zupan. 1996. Commuter and light rail transit corridors: The land use connection. Transit Cooperative Research Program 16, volume 1, March.

Chartered Institution of Highways & Transportation. 1995. *Developing Urban Transport Strategies*. London, England: CIHT. http://www.ciht.org.uk/en/knowledge/publications/technical-guidelines.cfm/developing-urban-transport-strategies-1995.

Coccia, E., P. D. Site, F. Filippi, M. Lemessi and A. Mallamo. 1999. Design of passenger interchanges. *International Conference Series on Competition and Ownership in Land Passenger Transport—1999, Thredbo 6*. Cape Town, South Africa. http://ses.library.usyd.edu.au:80/handle/2123/6510.

Collet, R. and T. Kuhnimhof. 2008. Relevant market segments in intermodal passenger travel. Deliverable D4, KITE project. http://www.kite-project.eu/kite/cms/index.php?option=com_content&task=blogcategory&id=12&Itemid=22.

Commission for Integrated Transport. 2000. *Physical Integration*. London, England: CfIT. http://webarchive.nationalarchives.gov.uk/20110304132839/http://cfit.independent.gov.uk/pubs/2000/physical/physical/index.htm (accessed January 2013).

Consorcio Regional de Transportes de Madrid. 2010. Plan de Intercambiadores—Madrid. Madrid: Consejería de Transportes, Infraestructuras y Vivienda (CRTM).

Cré, I., S. Bührmann, S. Edwards, D. Jeffery, J. Monigl and A. Szekely. 2008. Selection of innovative concepts. Deliverable D1.2, NICHES + project. http://www.transport-research.info/web/projects/project_details.cfm?id=11075.

Crockett, J., M. Beecroft, M. McDonald, T. Whiteing, G. Whelan, C. Nash, T. Fowkes and M. Wardman. 2004. Rail user needs and delivery mechanisms. Rail Research UK report RRUK/C2/02.

Crozet, Y. and I. Joly. 2004. Budgets temps de transport: Les sociétés tertiaires confrontées à la gestion paradoxale du 'bien le plus rare'. *Les Cahiers Scientifiques du Transport* 45: 27–48.

Dell'Asin, G., M. E. Lopez-Lambas and A. Monzon. 2014. Key quality factors at urban interchanges. *Proceedings of the ICE—Transport*, Mayo, pp. 1–10.

dell'Olio, L., A. Ibeas and P. Cecin. 2011. The quality of service desired by public transport users. *Transport Policy* 18(1): 217–227.

de Oña, J., R. de Oña and F. J. Calvo. 2012. A classification tree approach to identify key factors of transit service quality. *Expert Systems with Applications*
39(12): 11164–11171.

Diaz, S. E., J. M. de Urena and C. Ribalaygua. 2012. Transport interchanges
effects on their surroundings in Tunja (Colombia) and Cordoba (Spain): A
comparative approach. *The Open Geography Journal* 5(1): 38–47.

Di Ciommo, F. 2002. L'accessibilité: L'enjeu prioritaire de la nouvelle politique des
transports publics à Naples. In B. Jouve (sous la direction de), *Les politiques
de déplacements urbains en Europe: L'innovation en question dans 5 villes
europeennes,* Paris, L'Harmattan, pp. 135–159.

Di Ciommo, F. 2004. La regénération urbaine à Naples, Paris et Milan: la fiabil-
ité du politique, condition de participation des acteurs économiques. PhD.
Thesis. Marne-la-Vallée, ENPC. http://www.theses.fr/2004ENPC0443.

Di Ciommo, F. and K. Lucas. 2014. Evaluating the equity effects of road-pricing
in the European urban context—The Madrid metropolitan area. *Applied
Geography* 54: 74–82.

Di Ciommo, F., A. Monzon, R. de Oña, J. de Oña and S. Hernandez. 2014. Using
hybrid latent class model for City-HUBs' users behaviour analysis. *XI Congreso
de Ingeniería del Transporte*, CIT 2014, 9–11 June, Santander, Spain.

Di Ciommo, F., J. M. Vassallo and A. Oliver. 2009. Private funding of intermodal
exchange stations in urban areas. *Transportation Research Record: Journal
of the Transportation Research Board* 2115(12): 20–26.

Dufour, D. 2010. PRESTO Cycling Policy Guide: Cycling Infrastructure. PRESTO
project—Promoting cycling for everyone as a daily transport mode. http://
ec.europa.eu/energy/intelligent/projects/en/projects/presto.

Durmisevic, S. and S. Sariyildiz. 2001. A systematic quality assessment of underground spaces—Public transport stations. *Cities* 18(1): 13–23.

Eckstein, H. 1975. Case study and theory in political science. In F. I. Greenstein
and N. W. Poisby (Eds), *Handbook of Political Science*, vol. VII, pp. 79–137.
Reading, MA: Addison-Wesley.

Edwards, B. 2011. *Sustainability and the Design of Transport Interchanges.*
Abingdon, UK: Routledge.

EITR, 2012. Urban mobility systems: Regulation across modes. In M. Finger,
I. Branf-Weiner, M. Holterman, and A. Russo (Eds), *1st European Intermodal
Transport Regulation Forum.* Fiesole, 7 December 2012. http://fsr.eui.eu/
Publications/WORKSHOPPAPERS/Transport/2012/121207EIntermodalTRS.
aspx.

European Commission. 2001. White Paper. European transport policy for 2010:
Time to decide. COM (2001) 370 final. Brussels.

European Commission. 2004. Analysis of the national inventories on passenger intermodality. Towards passenger intermodality in the EU. Report 2
Dortmund, October 2004. http://trid.trb.org/view.aspx?id=761986.

European Commission. 2006. Keep Europe moving: Sustainable mobility for our
continent. Mid-term review of the European Commission's 2001 Transport
White Paper. Luxemburg: European Commission.

European Commission. 2007. Green Paper: Towards a new culture for urban
mobility. COM (2007) 551 final. Brussels.

European Commission. 2008. Action plan for the deployment of intelligent transport systems in Europe. COM (2008) 886 final. Brussels.

European Commission. 2009. Action plan on urban mobility. COM (2009) 490/5. Brussels.

European Commission. 2011. White Paper. Roadmap to a single European transport Area—Towards a competitive and resource-efficient transport system. COM (2011) 144 final. Brussels.

European Commission. 2013a. A concept for sustainable urban mobility plans. Annex 1 to COM (2013) 913 final. Brussels.

European Commission. 2013b. Thematic research summary: Passenger transport. http://www.transport-research.info/Upload/Documents/201302/20130227_131922_22068_TRS05.pdf.

European Union. 2011. Cities of tomorrow: Challenges, visions, ways forward. Publications Office of the European Union.

Ewing, R., T. Schmid, R. Killingsworth, A. Zlot and S. Raudenbush. 2008. Relationship between urban sprawl and physical activity, obesity, and morbidity. In J. M. Marzluff, E. Shulenberger, W. Endlicher, M. Alberti, G. Bradley, C. Ryan, U. Simon and C. ZumBrunnen (Eds), *Urban Ecology*, pp. 567–582. New York: Springer.

Field, A. 2009. *Discovering Statistics Using SPSS: (and Sex, Drugs and Rock'n'Roll)*. 3rd edn, New York: Sage.

Filion, P. and A. Kramer. 2012. Transformative metropolitan development models in large Canadian urban areas: The predominance of nodes. *Urban Studies* 49(10): 2237–2264.

Geurs, K. T., W. Boon and B. Van Wee. 2009. Social impacts of transport: Literature review and the state of the practice of transport appraisal in the Netherlands and the United Kingdom. *Transport Reviews* 29(1): 69–90.

Grafl, W., G. Huber, G. Sammer, J. Stark, T. Uhlmann, W. Unbehaun and S. Wegener. 2008. Catalogue of best-practice implementation examples. Deliverable D13, KITE project. http://www.kite-project.eu/kite/cms/index.php?option=com_content&task=blogcategory&id=12&Itemid=22.

Graham-Rowe, E., S. Skippon, B. Gardner and C. Abraham. 2011. Can we reduce car use and, if so, how? A review of available evidence. *Transportation Research Part A: Policy and Practice* 45(5): 401–418.

Green, C. and P. Hall. 2009. Better rail stations. London Department for Transport Great Minster House. http://collections.europarchive.org/tna/20100409091328/http:/www.dft.gov.uk/pgr/rail/passenger/stations/betterrailstations/.

Grémy, J-P. and M-J. Le Moan. 1977. Analyse de la démarche de construction de typologies dans les sciences sociales. *Informatique et Sciences Humaines* 35.

Grotenhuis, J-W., B. W. Wiegmans and P. Rietveld. 2007. The desired quality of integrated multimodal travel information in public transport: Customer needs for time and effort savings. *Transport Policy* 14(1): 27–38.

Guo, Z. 2008. Transfers and path choice in urban public transport systems. PhD. Thesis. Transportation Department, Massachusetts Institute of Technology, Cambridge, MA. http://dspace.mit.edu/handle/1721.1/45401.

Guo, Z. and N. H. M. Wilson. 2011. Assessing the cost of transfer inconvenience in public transport systems: A case study of the London Underground. *Transportation Research Part A: Policy and Practice* 45(2): 91–104.

Harmer, C., K. Millard, D. Palmer, B. Ubbels, A. Monzon and S. Hernandez. 2014. What makes a successful urban interchange?: Results from an evidence review. *Transport Research Arena (TRA) 5th Conference: Transport Solutions from Research to Deployment—Innovate Mobility, Mobilise Innovation*. Paris, France. http://trid.trb.org/view.aspx?id=1327834.

Heddebaut, O. and D. Palmer. 2014. Multimodal city-hubs and their impact on economic and land use planning. In *TRA2014 Transport Research Arena 2014. Transport Solutions: From Research to Deployment—Innovate Mobility, Mobilise Innovation*. Paris, France, pp. 14–17.

Hermes. 2011. Interoperability barriers to intermodality and interconnectivity of passenger transport. Deliverable 4.I, HERMES—High Efficient and Reliable Arrangements for Crossmodal Transport. http://www.transport-research. info/Upload/Documents/201306/20130628_164614_73799_presentation_ BarriersToIntermodalityInterconnectivityOfPassengerTransport.pdf.

Hernandez, S. and A. Monzon. 2016. Key factors for defining an efficient urban transport interchange: Users' perceptions. Cities 50, 158–167.

Hernandez, S., A. Monzon and R. de Oña. 2015. Urban transport interchanges: A methodology for evaluating perceived quality. *Transportation Research Part A: Policy and Practice* 84: 31–43.

Hine, J. and J. Scott. 2000. Seamless, accessible travel: Users' views of the public transport journey and interchange. *Transport Policy* 7(3): 217–226.

Iseki, H. and B. Taylor. 2010. Style versus service? An analysis of user perceptions of Transit Stops and Stations. *Journal of Public Transportation* 13(3), 23–48.

Kashani, A. T. and A. S. Mohaymany. 2011. Analysis of the traffic injury severity on two-lane, two-way rural roads based on classification tree models. *Safety Science* 49(10): 1314–1320.

Levinson, D. 2010. Equity effects of road pricing: A review. *Transport Reviews* 30(1): 33–57.

Lijphart, A. 1971. Comparative politics and the comparative method. *American Political Science Review* 65(3): 682–693.

Litman, T. 2012. Evaluating non-motorized transportation benefits and costs. Canada: Victoria Transport Policy Institute. https://cppwbe.files.wordpress. com/2011/08/evaluating-non-motorized-benefits-and-costs-may-2011-vtpi. pdf.

Liu, R., R. Pendyala and S. Polzin. 1997. Assessment of intermodal transfer penalties using stated preference data. *Transportation Research Record: Journal of the Transportation Research Board* 1607(1): 74–80.

Lucas, K. and P. Jones. 2009. *The Car in British Society*. London: RAC Foundation. http://www.racfoundation.org/assets/rac_foundation/content/downloada- bles/car_in_british_society-lucas_et_al-170409.pdf.

Lyons, G. and J. Urry. 2005. Travel time use in the information age. *Transportation Research Part A: Policy and Practice* 39(2–3): 257–276.

Macário, R. and E. Van de Voorde. 2010. *Critical Issues in Air Transport Economics and Business*. London: Routledge.

Mackie, P. J., S. Jara-Díaz and A. S. Fowkes. 2001. The value of travel time savings in evaluation. *Transportation Research Part E: Logistics and Transportation Review* 37(2–3): 91–106.

Mannone, V. 1997. Gares TGV et nouvelles dynamiques urbaines en centre-ville: Lecas des villes desservies par le TGV Sud-Est. *Les Cahiers Scientifiques du Transport*, n°31/1997: 71–97.

Martens, K. 2007. In G. De Roo and G. Porter (Eds), Actors in a fuzzy governance environment. *Fuzzy Planning: The Role of Actors in a Fuzzy Governance Environment*, Burlington, VT: Ashgate.

Martilla, J. A. and J. C. James. 1977. Importance-performance analysis. *Journal of Marketing* 41(1): 77–79.

Metz, D. 2008. The myth of travel time saving. *Transport Reviews* 28(3): 321–336.

Monzon, A., A. Alonso and M. E. Lopez-Lambas. 2013. Key factors affecting the efficiency of transport interchanges. *13th World Conference on Transport Research (WCTR)*. Rio de Janeiro, Brazil. http://oa.upm.es/30239/.

Morris, M., M. Schindehutte and J. Allen. 2005. The entrepreneur's business model: Toward a unified perspective. *Journal of Business Research*, Special Section: The Nonprofit Marketing Landscape, 58(6): 726–735.

Nathanail, E. 2008. Measuring the quality of service for passengers on the Hellenic railways. *Transportation Research Part A: Policy and Practice* 42(1): 48–66.

Naude S., J. Joner and P. Louw. 2005. Design guidelines for public transport facilities. SATC 2005. http://repository.up.ac.za/bitstream/handle/2263/6332/041. pdf?sequence=1.

Navarrete, F. J. and J. D. Ortuzar. 2013. Subjective valuation of the transit transfer experience: The case of Santiago de Chile. *Transport Policy* 25, 138–147.

Network Rail. 2011. *Guide to Station Planning and Design*. Issue 1, Network Rail, London.

Offner, J-M. 1993. Les 'effets structurants' du transport: Mythe politique, mystification scientifique. *Espace Géographique* 22(3): 233–242.

Osterwalder, A. 2004. The business model ontology: A proposition in a design science approach. PhD. Thesis. Université de Lausanne, Lausanne, Switzerland. http://www.hec.unil.ch/aosterwa/PhD/Osterwalder_PhD_BM_Ontology.pdf.

Osterwalder, A., Y. Pigneur and C. L. Tucci. 2005. Clarifying business models: Origins, present, and future of the concept. *Communications of the Association for Information Systems*. Vol. 16. University of Lausanne, Switzerland. http://aisel.aisnet.org/cais/vol16/iss1/1.

Passenger Focus. 2011. The challenge of getting to the station: Passenger experiences. http://www.transportfocus.org.uk/research/publications/the-challenge-of-getting-to-the-station-passenger-experiences.

Peek, G-J. and M. van Hagen. 2002. Creating synergy in and around stations: Three strategies for adding value. *Transportation Research Record: Journal of the Transportation Research Board* 1793(1): 1–6.

Pirate. 2001. Final report. PIRATE: Promoting Interchange Rationale, Accessibility and Transfer Efficiency. Sheffield. http://www.transport-research.info/ Upload/Documents/200310/pirate.pdf.

Pitsiava-Latinopoulou, M. and P. Iordanopoulos. 2012. Intermodal passengers terminals: Design standards for better level of service. *Procedia—Social and Behavioral Sciences*, Transport Research Arena 2012 48: 3297–3306.

Plassard, F. 1977. Les autoroutes et le developpement regional. *Economie appliqué* 1: 225–244.

Pressman, J-L. and A. B. Wildavsky. 1973. *Implementation*. Berkeley, CA: University of California Press.

Preston, J. 2001. Integrating transport with socio-economic activity—A research agenda for the new millennium. *Journal of Transport Geography* 9(1): 13–24.

Preston, J., G. Wall and T. Whiteing. 2006. Delivery of user needs: Final report. Rail Research UK report: RRUK/C2/06. http://www.researchgate.net/publication/238667458_Prepared_for_Rail_Research_UK.

Ragin, C. C., D. Berg-Schlosser and G. De Meur. 1996. Political methodology: Qualitative methods. In *A New Handbook of Political Science*: New York: Oxford University Press 749–768.

Rail Safety and Standards Board. 2010. Topic note on integrated transport (T824). RSSB. http://www.rssb.co.uk/research-development-and-innovation/research-and-development/research-project-catalogue/t824.

Rail Safety and Standards Board. 2013. Guidance on the implementation of station travel plans. RSSB. http://www.researchgate.net/publication/273259008_Guidance_on_the_implementation_of_Station_Travel_Plans.

Richer, C. 2008. L'émergence de la notion de pôle d'échanges, entre interconnexion des réseaux et structuration des territoires. *Les Cahiers scientifiques du transport* 54(12): 101–123.

Riley, P., A. Kumpostova, R. Tommasi, J. F. Collet, C. Rogge, S. Haon, K. Vancluysen, S. Buhrmann, W. Backhaus and P. Hoenninger. 2010. Recommendations and strategies for passenger intermodality in Europe. Deliverable D.4.2., LINK project. http://www.transport-research.info/Upload/Documents/201302/20130205_142346_41929_LINK_recommendations_fullversion_4-2010.pdf.

Ryan, S. 1999. Property values and transportation facilities: Finding the transportation-land use connection. *Journal of Planning Literature* 13(4): 412–427.

Simon, O. 1996. The economic impact of transport infrastructure provision: A review of the evidence. *Proceedings of the Institution of Civil Engineers-Transport—PROC INST CIVIL ENGTRANSPORT.* 01/1996, 117(4): 241–247.

Sintropher. 2012. Good practice in transport interchanges: A range of case studies from across the northwest Europe area examining public transport interchange design. Project report. http://sintropher.eu/sites/default/files/images/editors/Project_Reports/Interchanges%20report%20web%20version%20updated.pdf.

Sputnic. 2009. Guidelines in market organisation—Public transport integration. Sputnic project—Strategies for public transport in cities. http://documents.rec.org/publications/SPUTNIC2MO_ptintegration_AUG2009_ENG.pdf.

Taylor, D. and H. Mahmassani. 1996. Analysis of stated preferences for intermodal bicycle-transit interfaces. *Transportation Research Record: Journal of the Transportation Research Board* 1556(1): 86–95.

Terzis, G. and A. Last. 2000. Urban interchanges—A good practice guide. GUIDE, Final Report. http://www.transport-research.info/Upload/Documents/200310/guide.pdf.

Transport for London. 2009. *Interchange Best Practice Guidelines*. London, England: Transport for London.

TRI-Value. 2014. TRI-Value: Ex-post evaluation of transport R&I in the FP7 Cooperation Programme Consortium. Final Report, European Commission.

Tsami, M., G. Adamos and E. Nathanail. 2013a. Investigating the accessibility level for disabled users at urban transport interchanges. *Proceedings of the 13th International Conference Reliability and Statistics in Transportation and Communication (RelStat'13)*, Riga, Latvia, pp. 109–116.

Tsami, M., G. Adamos and E. Nathanail. 2013b. Sustainable development for the design for the transformation of the Thessaloniki railway station into a City Hub. European Transport Conference (ETC) 2013, Frankfurt, Germany, http://abstracts.aetransport.org/paper/index/id/171/confid/1.

Tunbridge Wells Borough Council. 2009. Guide to engaging hard-to-reach groups. http://www.tunbridgewells.gov.uk/__data/assets/pdf_file/0005/13694/FINAL-AGS-200910.pdf.

van Hagen, M. 2011. *Waiting Experience at Train Stations*. The Netherlands: Eburon Academic.

van Hagen, M. and R. Martijnse. 2010. Resource pack: Railway station design. Trendy Travel—Emotions for Sustainable Transport. http://www.eltis.org/sites/eltis/files/trainingmaterials/resource_pack_for_station_design_en.pdf.

van Hagen, M. and G. J. Peek. 2003. What you want, is what you should get: Customers' wishes in relation to the redevelopment of inner-city railway station areas. Frankfurt, Germany: European Transport Conference. http://abstracts.aetransport.org/paper/index/id/1745/confid/9.

Vassallo, J. M., F. Di Ciommo and A. Garcia. 2012. Intermodal exchange stations in the city of Madrid. *Transportation* 39(5): 975–995.

Vuchic, V. R. 2005. *Urban Transit: Operations, Planning, and Economics*. New York: John Wiley.

Wang, Y., A. Monzon and F. Di Ciommo. 2015. Assessing the accessibility impact of transport policy by a land-use and transport interaction model–The case of Madrid. *Computers, Environment and Urban Systems* 49: 126–135.

Wardman, M. and J. Hine. 2000. Costs of interchange: A review of the literature. Institute of Transport Studies, University of Leeds, Working Paper 546. http://eprints.whiterose.ac.uk/2075/.

Wefering, F., S. Rupprecht, S. Buhrmann and S. Bohler-Baedeker. 2013. Guidelines. Developing and implementing a sustainable urban mobility plan. Germany: Rupprecht Consult. http://www.sustainable-urban-mobility-plans.org/docs/SUMP_guidelines.doc.

Woldeamanuel, M. G. and R. Cyganski. 2011. Factors affecting travellers' satisfaction with accessibility to public transportation. European Transport Conference. Glasgow, http://www.worldtransitresearch.info/research/4404/.

Yin, R. K. 2009. *Case Study Research: Design and Methods*. 4th edn, Applied social research methods series, vol. 5. New York: Sage.

Appendix I: Factsheets from case studies

CASE STUDY 1: MONCLOA, MADRID (SPAIN)

Location	Northwest limit of the city
Year developed	1995
Last refurbishment	2008
Passengers per day	287,000
Modes of transport	Local and regional buses and metro

Role of interchange

The Moncloa Interchange (Figure A1.1), situated at the north-western limit of Madrid, but in a built-up area, provides a gateway to the city for over 287,000 people per day. The 'Madrid Interchanges Plan' addresses the linking process between the exchange points of the metropolitan bus lines and the metro Circular Line. This plan consists of improvements to the existing transport interchange stations – Moncloa, Avenida de América and Príncipe Pío – and the construction of new interchange stations in Conde Casal, Legazpi, Chamartín and Plaza Elíptica in order to achieve a modal interchange network organised around Madrid's entrances in relation to the highways and the interior circular metro network.

Due to capacity problems resulting from the growth in demand within the northwest motorway corridor (A-6), the expansion of the Moncloa transport interchange station was carried out to relocate the metro Line 3 station to improve passenger transit and provide parking spaces for bus inspections, together with new installations and equipment. This development was essential in order to carry out the expansion of the station (Arch Module), as it freed up the required space that had previously been used for the metro Line 3 station and its garages.

Figure A1.1 Moncloa Interchange: entrance (a); exclusive direct access to the interchange for buses from the highway (b); bus island (c); metro (d). (Courtesy of Jan Vasicek.)

Location

The interchange is located at an entrance point to Madrid surrounded by many historic monuments. It connects directly to metro Line 6, the Circular Line, which travels around the centre of the city and links with all of the key points on the metro network. The opening of this station has achieved excellent results, not only in terms of increased demand, but also in reductions in surface-level bus stops and improved journey times for both users and transport companies.

Modes of transport

There are currently 56 regional bus routes, with over 4000 journeys per day, and 310 journeys per hour between 8 and 10 a.m. Bus services in the peak hour are every 5–10 min, and they access the underground bus station by using a high-occupancy vehicle (HOV) lane. Apart from the 56 regional bus lines, there are 20 urban bus lines, 2 metro lines (Line 3 and Line 6) and 1 long-distance bus line. Moncloa is the metro station with the highest daily demand in Madrid. No private car parking is provided.

A total of 12 operators gather in Moncloa Interchange station (9 for regional bus, 1 for urban bus, 1 for long-distance bus and 1 for metro).

CASE STUDY 2: KAMPPI, HELSINKI (FINLAND)

Location	City centre
Year developed	2005
Last refurbishment	—
Passengers per day	84,000
Modes of transport	Local and long-distance buses, metro, rail, tram, cycle, taxi and private car

Role of interchange

Kamppi terminal in Helsinki (Figure A1.2) comprises three separate terminals: a terminal for local western buses with about 1,000 buses a day, a metro station underneath and a separate terminal for long-distance buses. Outside the terminal, there is on-street access to trams and inner city buses in the immediate vicinity. In addition, Kamppi terminal is a part of the city interchange area, with direct access to the main railway station, which is the hub for all local railway lines as well as for long-distance trains, and for

Figure A1.2 Kamppi Interchange: entrance with cycle parking (a); bus bays (b); metro (c); retail (d). (Courtesy of Jan Spousta.)

the two smaller bus terminals for northern and eastern buses. The terminal is situated in the basement of the newest shopping centre 'Kamppi' in the inner city of Helsinki. Kamppi has been designed by Finnish architect Juhani Pallasmaa, and it was opened in 2005.

Location

The interchange itself is in a very central location in downtown Helsinki. The area of the interchange/shopping centre is approximately 4 hectares. In addition to the shopping centre, there are also offices and flats in the same building complex. At the time when Kamppi was opened, the population of the city of Helsinki was ca. 561,000, nearly 1 million were living in the Helsinki metropolitan area and a total of 1.4 million people were living in the entire commuting area (25% of Finland's population). In addition, Helsinki has 9 million national and 2 million international tourists yearly.

Modes of transport

Modes of transport at the interchange include local, regional, national and international buses (to St. Petersburg, Russia), metro, tram, bicycle, car and taxi. The average number of visitors to Kamppi on working days is approximately 100,000, of which 84,000 use public transportation. The average number of departing metro passengers from Kamppi for all working days of the year is 21,500. The average number of bus passengers departing on working days from the local (western) terminal in Kamppi is 19,500.

The central railway station is approximately 500 m away from the Kamppi Interchange and can be reached by either an underground walking tunnel or the metro. Located adjacent to the railway station, there are bus stations for most of the northern and eastern local and regional buses.

Parking at Kamppi is located underground the terminals and the shopping centre. P-City and P-Kamppi parking garages offer 750 parking spots with access from all directions. For cyclists, Kamppi has a bicycle centre offering parking facilities, rental and maintenance services.

CASE STUDY 3: ILFORD RAILWAY STATION, LONDON (UNITED KINGDOM)

Location	Suburban town centre
Year developed	1839
Last refurbishment	1980s
Passengers per day	21,000 (rail only)
Modes of transport	Main-line rail, bus, cycle (with cycle parking), private car with drop-off, car parking and taxi

Role of interchange

Ilford Railway Station (Figure A1.3) is situated on the Great Eastern Main Line and has regular local train services (from Essex) to Liverpool Street Station in central London. Nearby bus stops are located within walking distance of the station, with the town being a hub of the London Buses network, providing buses to central London and various suburbs. The station is considered to be a major public transport interchange by Transport for London (TfL).

The interchange is planned for redevelopment as part of the Crossrail project.* The existing station is to be reconfigured to serve Crossrail trains from 2019. This will provide more than twice the current frequency of trains from Ilford to central London and is expected to encourage significant increases in passenger numbers. The station improvements will provide a new ticket hall layout with greater gate line capacity, passenger lifts, longer platforms and a realigned station entrance and elevation to the street.

Location

Ilford is a large suburban town in the London Borough of Redbridge, East London. The town is a significant commercial and shopping district surrounded by extensive residential development. Redbridge is an outer London Borough with a population of almost 300,000, having grown rapidly in the early twentieth century as a residential area serving as a satellite to central London. According to Redbridge Council statistics, Redbridge is the ninth most diverse borough in the country with approximately 55% of its population coming from a minority ethnic background.† Also, Redbridge was ranked as the 134th most deprived borough in the

---

* Crossrail is a railway, 118 km in length, currently under construction that will link Maidenhead and Heathrow Airport to the west of London with Shenfield and Abbey Wood to the east via Greater London with 42 km of new tunnels and new underground stations in central London.
† http://www2.redbridge.gov.uk/cms/the_council/about_the_council/about_red-bridge/2011_census/diversity.aspx.

(a)

(b)   (c)

Figure A1.3 Ilford Railway Station: entrance (a); cycle parking at the rail platform (b); rail platforms (c). (Courtesy of Derek Palmer (a) and Jan Spousta (b and c).)

country (out of 326). The Valentines area of the Borough – which contains Ilford Station and the main shopping areas – is in the highest 10% band of deprivation.*

The town centre, in which the interchange is located, is dominated by a heavily trafficked gyratory road system (A118) around Chapel Road, Ilford

---

* Deprivation in Redbridge Report 2010. http://www2.redbridge.gov.uk/cms/the_council/about_the_council/about_redbridge/research_and_statistics/deprivation_in_redbridge.aspx.

Hill and Cranbrook Road; this connects to the nearby North Circular Road (A406), which provides a key orbital link around London.

Modes of transport

This interchange is on the Great Eastern Main Line and has regular local train services from Essex to Liverpool Street Station. The station has five platforms, two for trains into London (towards Liverpool Street) and two out of London (towards Shenfield). The modes of transport available at the interchange are: main-line rail, bus, cycle (with cycle parking), private car with drop-off, car parking and taxi. More than 10 bus stops are nearby, with the town being a hub of the London Buses network. The station is located within TfL Zone 4.* Most trains stopping at Ilford run between Shenfield and London Liverpool Street, with at least 6 trains per hour train in each direction. Train services are currently within the Greater Anglia rail franchise, operated by Abellio Greater Anglia.

Most interchanges at Ilford will be local to local, with some being local to regional, for example, for passengers who travel out to Southend.

---

* London Underground, Docklands Light Railway (DLR), London Overground and National Rail services in London are divided into zones. Most services operate in Zones 1–6, with London Underground, London Overground and National Rail also operating in Zones 7–9.

CASE STUDY 4: KÖBÁNYA-KISPEST, BUDAPEST (HUNGARY)

Location	Suburban
Year developed	1978–1980
Last refurbishment	2008–2011
Passengers per day	155,500
Modes of transport	Rail, metro, local and regional buses, private car and taxi

Role of interchange

Köbánya-Kispest (Figure A1.4) is the terminal of metro line M3, which is one of the backbones of public transport in Budapest connecting the northeast and southeast of the city via the city centre. Köbánya-Kispest is also a major railway station, but most trains do not terminate there.

Local and regional bus connections are provided from a new bus terminal under the shopping mall, which has been constructed as part of a general refurbishment of the area. Although all types of connections are available, the interchange handles primarily local and suburban traffic.

Location

The interchange divides two areas of the city that are very different in character. North of the railway tracks, the area is a mixture of industrial sites, with a large number of abandoned factory buildings and a large, densely built high-rise housing estate (Újhegy) with prefabricated concrete buildings built in the 1970s. South of the terminal, there is another large housing estate (Kispesti lakótelep) and detached houses.

Modes of transport

There are several possibilities for interchange between transport modes in the intermodal centre, including railway, metro, local and regional buses, as well as walking, cycling and cars.

Köbánya-Kispest is the terminal of metro line M3, which connects the northeast and southeast of the city via the city centre. It has a nominal capacity of 28,200 passengers/hour/direction and approx. 630,000 passengers a day. The highest frequency is every 2.5 min during peak hours.

Köbánya-Kispest is a major railway station on railway lines No. 100 (Budapest – Cegléd – Szolnok – Debrecen – Záhony; suburban, regional, intercity and international services) and No. 142 (Budapest – Lajosmizse; suburban services only). Most trains do not terminate at Köbánya-Kispest but in Budapest-Nyugati, a major railway station in the city centre.

(a)

(b)

(c)

Figure A1.4 Köbánya-Kispest Interchange: access zone (a); bus stop under the shopping mall (b); rail platforms and footbridge (c). (Courtesy of Andres Monzon (a) and Andres Garcia (b and c).)

Local bus connections are provided by BKV, the local transport provider in Budapest, from the new bus terminal under the shopping mall. Three suburban lines to the eastern suburbs also terminate at the bus terminal operated by the regional operator Volánbusz. Terminal 2 of Liszt Ferenc International Airport is linked to the terminal by an express bus (200E) with 140 departures per day to the airport; the travel time is about 30 min.

Four parking lots are available: two park-and-ride facilities – one covered and one open air; a three-storey parking garage at the mall; and an open-air parking lot at the local hardware store (OBI). There is no dedicated kiss-and-ride zone. Bicycle parking is provided for cyclists.

CASE STUDY 5: RAILWAY STATION OF THESSALONIKI, THESSALONIKI (GREECE)

Location	Urban
Year developed	1961
Last refurbishment	1998
Passengers per day	166,500
Modes of transport	Rail, local, suburban and interurban buses, taxi and private car (metro under construction)

Role of interchange

The Railway Station of Thessaloniki (Figure A1.5) accommodates railway passengers who travel between the city and the suburban area, as well as other national and international destinations. Thessaloniki is the second biggest city in the country and the capital of the prefecture of Central

(a)   (c)

(b)   (d)

Figure A1.5 Railway Station of Thessaloniki: entrance to the building (a); ticket selling and waiting area (b); bus terminal in front of the interchange building (c); rail platforms (d). (Courtesy of Natalia Sobrino (a and b), Jan Vasicek (c) and Jan Spousta (d).)

Macedonia. The station is located in the urban area of the city, close to the 'Western Exit' Highway.

Apart from the railway, the station serves as one of the two bus terminals in the city, and accommodates urban and interurban buses. The new metro network's terminal, currently being constructed, is also located at the New Railway Station of Thessaloniki. There will also be new underground parking and new walking and cycling facilities. The project is expected to highly affect the surrounding area, providing more incentives for new businesses, attracting housing relocation and increasing land and property value in the area.

Location

The station is situated very close to the central business district, allowing the movement of travellers all around the city. The station is also close to the port of Thessaloniki, enhancing the attractiveness of the interchange. Moreover, there is a bus line connecting the railway station with the Central Interurban Bus Terminal located in the west part of the city and with the International Airport of Thessaloniki, 'Macedonia', located in the east part.

Modes of transport

The station is located in the west part of the city and accommodates suburban and interurban rail, urban, suburban and interurban buses, taxis, bicycles, park-and-ride, kiss-and-ride and metro (under construction). The existing terminal building, in combination with the investment in the new metro station, will enable the development of the existing infrastructure into a modern integrated bus–railway–metro station.

Located at the station is one of the two city terminals for the urban bus services of the Thessaloniki Urban Transport Organisation (OASTH). The station is also directly connected to the interurban bus station, where there are routes to Athens and other Greek cities.

The ridership that uses the station daily comprises of an average of 138,000 bus passengers travelling in the urban zone and 22,500 passengers travelling in the suburban zone, on a total of 12 bus lines. The average daily number of railway passengers arriving at/or departing from the station is approximately 6,000. Of these, 4,500 use tickets issued by electronic systems, and 1,500 use paper tickets.

CASE STUDY 6: GARES LILLE FLANDRES
AND LILLE EUROPE, LILLE (FRANCE)

Location	City centre
Year developed	1848 (Lille Flandres) 1994 (Lille Europe)
Last refurbishment	2008 (Lille Flandres metro station) and 2013–2015 (Lille Flandres) 2014 (Lille Europe)
Passengers per day	110,000 (Lille Flandres) 8,500 (Lille Europe) (rail only)
Modes of transport	Rail, high-speed rail, metro, tram, bus, car sharing, self-service bikes, taxis, bike taxis

Role of interchange

This case study is composed of two main interchanges in the Lille metropolis: Lille Flandres and Lille Europe (Figure A1.6). Effectively, they are located near each other at about 500 metres in the 'railway stations triangle' and offer possibilities to transfer from urban public transports (metro, tramway, buses, self-service bikes, car-sharing) to rail services at a mainly regional level (Lille Flandres) and mainly at national and international

(a)

(c)

(b)

(d)

Figure A1.6 Lille Flandres outside (a) and inside (b); Lille Europe outside (c) and inside (d). (Photographs copyright Gares et Connexions SNCF (a and b), and courtesy of Jan Vasicek (c) and Jan Spousta (d).)

levels (Lille Europe). A large shopping mall named Euralille is located between the two stations.

In 2008, a first refurbishment was made of the Lille Flandres metro and tram stations. The Lille Flandres railway station has been under refurbishment since 2013 for better train services, and from 2014 until 2015 to provide a new concourse and a new joint public transport and rail ticket office (purchase of rail or PT tickets, joint information, network maps, etc.). These works will also provide wider access to public transport linking the ground floor level to underground levels. The two first floors of the station will be converted into a new 1,300 m^2 business centre with commercial and office spaces. Every day, 70,000 passengers take the train.

Lille Europe is the new high-speed train station for TGV and Eurostar, Thalys and TER-GV (high-speed regional trains). The Lille Europe railway station is a very modern railway station constructed to host the northern TGV trains on the high-speed railway network. It opened in 1994 with the opening of the Channel Tunnel and the northern TGV network. It has a daily traffic of 8,500 passengers.

Location

Lille Flandres is the regional station for TER trains and TGV towards Paris railway station. The Lille Flandres railway station is an old construction station inside the city of Lille, very close to the main square of Lille, and near the old part of the city. This station opened in the nineteenth century in 1848. Its façade is the previous 'Gare du Nord' from Paris that was reconstructed in Lille in 1867.

Modes of transport

Lille Flandres counts 17 platforms for more than 500 trains per day and has a traffic of 20 million passengers per year (2012) and 110,000 daily users, of which 70,000 take the train and 40,000 are only crossing, using services or shopping.

The Lille Flandres railway station serves the regional towns with regional trains named TER (Express Regional Trains). It also links Lille to Paris in one hour with direct TGV trains. Lille Flandres has a lot of connections with the urban transport network with the Metro VAL Line 1 and Line 2 (located underground), and two tramway lines towards Roubaix and Tourcoing (at intermediate level). It is close to regular bus routes and one specific bus route, Citadine, that serves the two railway stations and a circle route inside Lille. It also recently offers self-service bikes in free access stations named V'Lille and a 'watch out' bike garage, which is free for public transport passengers and train users. The Lille Flandres metro station counts 170,000 passengers per day, of which about 48,400 are train passengers coming or going to the Lille Flandres railway station. At peak hour, there is a metro every minute at peak hour and every five minutes at off-peak hours.

Lille Europe counts 4 platforms with 2 central railway lines for passing Eurostar trains coming from Paris and going directly to London. It connects Lille to Paris by TGV (1 h), and to Brussels (38 min) and London (1 h 20 min) with the Eurostar trains. It also serves the other French regions by TGV trains that go directly to the south east (Lyon, Marseille, Nice), or the east (Strasbourg). It also connects the western parts of France (Brittany to Nantes, Rennes, and south west to Bordeaux) by means of trains that go around Paris and westwards afterwards. All the TGVs stop at the international airport of Roissy Charles De Gaulle when going to the south.

The Lille Europe passengers mainly come by car or metro. Since 2014, works have been undertaken on the departure concourse for the cross-Channel trains (Eurostar): they will provide new control desks, increasing the boarding area. It also offers new facilities such as Wi-bike (biking plugs), a piano, a children's area and so on.

The Lille Europe railway station is also well connected to urban public transport with the VAL metro Line 2 by means of a metro station next to the main concourse, the tramway Lines 1 and 2 by a tramway station close to the railway station, and regular urban buses. These two railway stations offer paid parking facilities.

CASE STUDY 7: UTRECHT CENTRAAL, UTRECHT (THE NETHERLANDS)

Location	City centre
Year developed	1843
Last refurbishment	2008 and ongoing
Passengers per day	285,000
Modes of transport	Rail, bus, *sneltram*, cycle (with cycle parking), private car (with car parking) and taxi

Role of interchange

Due to the central location of Utrecht Centraal (Figure A1.7) within the Dutch rail network, Utrecht Centraal is considered the most important railway hub of the Netherlands: trains from almost all directions pass through. Important intercity connections are, amongst others, the intercity Maastricht (Amsterdam and further), the intercity Nijmegen (Den Helder) and the North-East, which is a name for the intercity trains from the north and east of the Netherlands heading to Randstad.

Figure A1.7   Overview of Utrecht Centraal (planned). (Image copyright CU2030. Source: cu2030.nl.)

Location

The Utrecht Central Interchange, owned by ProRail, NS, is the central railway station of the city of Utrecht in the Netherlands. The station is centrally located in the country, being the largest station regarding passenger volume. More than 900 trains depart from the station per day and almost 285,000 passengers are connected daily to nearby areas or cities, other Randstad cities or international destinations.

Utrecht Centraal Station was built for about 35 million passengers per year. At the moment, around 88 million passengers use the station. This amount will grow in the future: the number of passengers expected in 2030 is about 100 million. The new terminal will become an easy-to-recognise and independent building. The current capacity is 284,000 users per day. The capacity in 2025 will be 360,000 users per day.

Between 2009 and 2016, Utrecht Centraal will be refurbished as a new OV-terminal. This new terminal is necessary due to the capacity limits and the expectation of ongoing growth. To be able to handle the expected future amount of passengers, several measures have been taken, including, amongst others, a terminal which integrates train, tram and bus. The renewal of the station is only one of the projects within the context of the project CU2030 in which plans for the complete station area of Utrecht are implemented.

Modes of transport

Besides being the most important railway hub of the Netherlands, Utrecht Centraal also has an extensive bus station. Utrecht Centraal has the biggest and busiest bus station in the Netherlands, and is an important interchange for city and regional bus services.

From the bus station CS Centralside, the city and regional bus lines from the Region Utrecht and Province of Utrecht depart from the eastern side of the rail. From the bus station CS Jaarbeurs, all bus lines in the direction of Zuid-Holland and Noord-Brabant depart, as well as lines departing in a western direction.

At the largest public transport interchange in the Netherlands, in the coming years, 22,000 attractive bicycle parking points will be built. Around the new Utrecht Centraal, over five large parking areas will be created on the western and eastern sides. Therefore, cyclists will be able to park as near as possible to the entrances of the station. Furthermore, in the area around the station, 11,000 bicycle parking facilities will be realised.

CASE STUDY 8: OSLO BUS TERMINAL VATERLAND, OSLO (NORWAY)

Location	City centre
Year developed	1989
Last refurbishment	2010
Passengers per day	27,500
Modes of transport	Local, regional, national and international buses, tram, metro, rail, taxi and private car

Role of interchange

The Oslo bus terminal Vaterland Interchange (Figure A1.8) is the main hub for regional, national and cross-border bus lines. In addition, it is a (relatively) short distance from and has good access to local buses, trams and subways that allow for further travel within or to the surrounding areas in Oslo. The bus station is about 400 m walking distance from the central railway station, which makes transfer between modes possible. Currently, there are plans to build a new bus terminal above the rail tracks and consequently reduce the distances between modes. There are also discussions connected to reducing the number of regional buses travelling to the city centre by making the buses travel to the more

(a)

(c)

(b)

(d)

Figure A1.8 Oslo bus terminal Vaterland: commuters transferring (a); bus lane and bus bays (b); info displays (c); bus bays area (d). (Courtesy of Julie Runde Krogstad (a and d) and Petter Christiansen (b and c).)

peripheral train and/or metro stations. The Oslo region experiences rapid population growth and it is expected that the growth will continue in the next decades. Population growth influences demand and the planning of regional public transport. The region has therefore recently conducted a plan that analyses the potential transport effects of the increased population. One main conclusion is that interchanges will become more important. There is limited capacity for buses driving into the Oslo city centre. Consequently, the authorities plan to establish transport services that to a larger extent transport travellers to regional interchanges.

Location

Oslo bus terminal is centrally located in the capital. The interchange is located within walking distance of the central business and trade districts in Oslo. Large shopping centres and shopping districts also lie next to the interchange. In recent years, there has been a rapid refurbishment and expansion in Bjørvika, located just south of the interchange. In 2016, the area will accommodate about 10,000 new jobs and approximately 500 new apartments. Such centralisation increases the importance and attractiveness of interchanges. Buses that operate at the interchange suffer from congestion in the nearby streets. This cause delays, especially during rush hours.

Modes of transport

Originally, the development of Oslo bus terminal was a collaboration between Akershus County and the municipality of Oslo. Consequently, Akershus County owns 78.5% of the terminal, while the municipality of Oslo owns 21.5%. The interchange was planned to receive about 450 daily departures and approximately 6,000 daily passengers. Since then, the capacity has been extended. In 2011, the number of daily departures were more than doubled. About 1,050 buses depart each day; 67% of the buses are operated through Ruter (the Public Transport Agency in Oslo and Akershus), 9% are airport express buses and 24% are coach. About 27,500 passengers pass through the interchange on an average day and about 10 million passengers pass through yearly.

CASE STUDY 9: PASEO DE GRACIA, BARCELONA (SPAIN)

Location	City centre
Year developed	1902
Last refurbishment	2014–2015
Passengers per day	100,000
Modes of transport	Rail and metro

Role of interchange

The Paseo de Gracia Interchange (Figure A1.9) is one of the most central stations in Barcelona. The trains operate within Catalonia, meaning that the train services are only possible for travel within the region and not to other parts of the country. The interchange also has three metro lines which serve more local travel. The metro lines run throughout the central areas in Barcelona as well as the suburbs. Next to the interchange is Paseo de Gracia, which is one of the most important shopping and business areas in Barcelona.

(a)   (b)

(c)   (d)

Figure A1.9 Paseo de Gracia Interchange: entrance ticket validation (a); ticket selling (b); people with cycles alighting from the metro (c); rail platform (d). (Courtesy of Lluis Alegre.)

Location

In the early twentieth century, when the railway line was being built, it was located at the most central point throughout the city, both then and now. The building was opened in 1902 and, despite having only two tracks and two platforms, it soon received as many passengers as the central station because of its privileged location.

It maintains, however, the great advantage of being extremely well located, as its catchment area has the highest density of jobs and amenities, which, despite its more defective design, leads to crowded stations.

Modes of transport

In 1924, Gran Metro, the first underground company in the city, built a subway line running under the Paseo de Gracia and perpendicularly crossing the railway line already in service; it has become L3 in the current underground network. Despite the proximity between metro and railway stations, and the fact that both stations were underground, no interchange corridor was foreseen between them. In the 1950s, the station became underground, as the corridor with the metro station was set up.

In the 1970s, once the first metro plan had been written and approved, one of the branches of Gran Metro was split up and became a part of the planned new L4. This necessitated the building of a new station on this line which, because of construction requirements, is located more than 250 m away from the existing L3 and, consequently, at a similar distance from the train station. A large neighbouring car park, with no relationship to the interchange, did not allow the construction of the lobby of the L4 station closer to L3. Consequently, the interchange corridor is unnecessarily long, to the point of becoming a deterrent. Moreover, it is the only one in the entire network between L3 and L4, one of the weak points in its functionality and connectivity.

In 1995, line L2 was opened, although its infrastructure had already been built in the 1960s. There is a station on this line named Paseo de Gracia as well, close to the L4 station of the same name, which allows for easy interchange. But for this same reason, the connection with L3 and the railway is extremely long and clearly also a deterrent.

The result today is that the Paseo de Gracia Interchange consists of three subway stations (L2, L3 and L4 metro lines) and a railway station on line R2. However, the distance from one end to the other exceeds 400 m, which in practice means some trips are too long.

CASE STUDY 10: PRAGUE TERMINUS
DEJVICKÁ, PRAGUE (CZECH REPUBLIC)

Location	Central area
Year developed	1920s/1978
Last refurbishment	2014
Passengers per day	150,000
Modes of transport	Metro, tram and bus

Role of interchange

The square has been an important transport hub since its foundation in the 1920s; however, its importance increased significantly with the opening of metro Line A in 1978. The new interchange incorporated a metro terminus, a bus service terminal and separate tram stops (Figure A1.10). The metro connects the district with the city centre (an extension to outskirt districts is under construction), while trams serve the neighbouring residential areas and connect the entire district with other parts of the city centre. Frequent local bus routes operate towards the Prague airport, the Suchdol district with the University of Agriculture campus and other residential districts, while several neighbouring municipal suburbs are served by bus routes with low frequency.

With the introduction of the integrated transport system around the year 2000, additional suburban routes serving municipalities adjacent to Prague were incorporated into the terminal (previously they terminated at other places). Since the opening of the metro, the terminal has also served as the interchange with bus services to Kladno (industrial city with population of 70,000, ca. 20 km out of Prague). The Kladno–Prague link is the busiest commuting flow in the country. The long-distance bus services have been added since the liberalisation of bus transport and access restrictions to the city centre.

Location

The Dejvická terminal is located at Vítězné náměstí (Victory Square), one of the biggest squares in Prague, and is a central point of the municipal district Prague 6. The entire area was designed, at the beginning of the area development in the 1920s–1930s, as a multipurpose district, incorporating major national, educational and district administrations and several types of residential zones.

Despite the importance of the area, its further development only began a few years ago. Several administrative buildings have been erected next to the metro entrance; however, the proposed completion of the square is still under discussion as it might disturb the entire character and sight of the square.

The entire area is suffering from road traffic congestion, serious environmental impacts and other disadvantages, due to the fact that this busy terminal is located in the middle of a residential district with very limited

(a)

(b)   (c)

Figure A1.10 Prague terminus Dejvická: panoramic view of Victory Square (a), tram stop (b) and bus stop (c). (Courtesy of Jan Vasicek.)

options for detours. The solution for the reduction of the local impact is seen in the metro. This goal has been partly reached by the metro Line A extension, which was opened on 6 April 2015 when relevant public transport services (buses from the airport, Kladno and Slaný) shortened to the new station Veleslavín. However, according to the first data available, the metro extension has had almost no impact on car traffic volumes.

Modes of transport

Travellers at the interchange can use the metro, trams and buses. The buses have a both a local and a regional network. The terminal serves more than 150,000 passengers per day; the metro station alone is used by around 120,000 passengers, making it one of the most frequently used stations in the network.

CASE STUDY II: INTERMODAL TERMINAL OF MISKOLC, MISKOLC (HUNGARY)

Location	City centre
Year developed	1901 (train station)
Last refurbishment	2014 train station, intermodal terminal is under development
Passengers per day	N/A
Modes of transport	Rail, metro, local and interurban buses, toll parking, taxi

Role of interchange

The station building was built in 1901 and refurbished in 2014 with respect to the functions and design of the original building. However, intermodal facilities or designs have not been implemented so far, and accessibility to the interchange remains difficult for pedestrians, cyclists and people with disabilities or with prams. In order to improve the interchange, notable efforts have been put into a large-scale project development with an intermodal approach. The aim of the planned project is to encourage people to favour public transport over other transport modes, by providing more adequate and reliable interchange conditions, improving the accessibility of the city, leading to a notable contribution to the reduction of environmental impacts. Thanks to this, public transport could become a competitive alternative to the car. One of the external benefits is the revitalisation of the surroundings (sport centre, thermal bath, retail).

Location

The terminal in Miskolc is located in one of the most populated cities in Hungary. The intermodal terminal is planned to be built as a refurbishment of the main train station of Miskolc, Tiszai Pályaudvar (Tiszai Railways Station), located in the western part of the city at Kandó Kálmán Square. Nowadays, this is the most important interchange in Miskolc.

Modes of transport

Currently, there are several transport modes available, such as train, tram (2 lines), local bus (5 lines), interurban bus (1 line), toll parking (50 places) and a taxi rank. The planned upgrades include an east-oriented head-end for the trams, to be replaced next to the local/interurban bus station. The interchange is expected to provide aesthetic, covered and spacious waiting areas and new parking lots, as well as new functions. The bus station is planned to be a landmark building in the square, and should function not only as a station but as a meeting point as well. Other functions, like P+R and B+R, are planned to be incorporated, supplemented with bike stands, separated walkways and the possibility of a future implementation of a tram–train solution. The arrivals and departures at the current Búza tér bus terminal (located in the city centre) are planned to be transferred into this new intermodal centre. Therefore, the new terminal had to be designed with respect to the need of serving the increasing heavy traffic as well.

The financial and management plan is roughly outlined in the study. According to this, the common areas will be operated jointly, which can be achieved either by founding a joint company or by a tender. The operation of other places is planned to be carried out by the municipality, while the maintenance costs are to be covered by different partners. According to the financial responsibilities presented in the study, the transportation providers will have the role of covering the costs of operational facilities. As for the facilities related to the supply of transportation, the actors that should cover the costs are the transportation providers and the renters of the commercial facilities. Regarding the rest, it is intended that the municipality will be responsible for the related payments.

CASE STUDY 12: INTERCITY COACHES OF MAGNESIA INTERCHANGE IN THE SUBURB OF VOLOS (GREECE)

Location	Suburban
Year developed	1970
Last refurbishment	1990
Passengers per day	2500
Modes of transport	Bus dominant

Role of interchange

The Intercity Coaches of Magnesia Interchange (Figure A1.11) was opened in the 1970s and an entire redevelopment took place in 1990s, including the refurbishment of the waiting and ticketing area, the storage area and the offices. The surrounding area of the interchange hosts university campuses, catering shops, coffee shops, restaurants and street markets, whereas housing development is not intense. The interchange has an ascendant role in the overall transport network, since it provides travelling services for 9 large cities (destinations out of Magnesia), 36

Figure A1.11 Magnesia Interchange: entrance to the building (a), bus operating zone (b and c). (Courtesy of Giannis Adamos.)

destinations in Magnesia and routes from and to the airport of Volos, 'Nea Anchialos National Airport'.

Location

The interchange is located at the west entrance of Volos, which is the capital of the Magnesia Prefecture, and adjacent to the local bus terminal. The railway station and the port are also close, at 1–1.5 and 1.5–2 km, respectively.

Although the interchange is very close to the university campuses, they are well segregated by a creek, so there is no interaction between the traffic generated by the campuses and the respective traffic generated by the interchange.

The access to the interchange is considered very good, and people may use the local buses or they can walk or cycle along the quays of the city's port to get there from the commercial/retail centre.

Modes of transport

The interchange mainly accommodates suburban and interurban buses. Other modes that the interchange caters for are local buses, cycling, walking, motorcycling, private cars with car parking and drop-off spots and taxis. The connectivity provided among the different modes at the terminal is considered adequate, since there is a connection between interurban and urban transit and taxis are available right outside the main building (10–15 m). Car and bicycle parking spaces are also available right outside the main building of the interchange (approximately 15–20 m).

The approximated average daily ridership of the interchange is comprised of 550 passengers travelling to the capital city of Greece, Athens; 400 passengers to Thessaloniki, which is the second largest city of Greece; and 500 passengers to other Greek cities (i.e. Larisa, Patra, Kozani, Ioannina, Agrinio, Lamia, Trikala). On average, around 1,200 trips have their destinations in Magnesia.

Appendix II: Interviews with practitioners

The City-HUB project conducted interviews with selected stakeholders related to interchanges in different countries across Europe in order to gain an understanding of the design of successful multimodal interchanges. This helped to understand the key factors that make an interchange effective from a variety of different viewpoints.

The interviews were held with practitioners (transport providers, local transport authorities or the business community) who were currently or had recently been involved in the design, management or operation of an urban interchange.

The interview form is reproduced here.

INTRODUCTORY INFORMATION

The City-HUB project, funded by the European Commission, is aimed at helping to design and operate seamless, smart, clean and safe intermodal public transport systems. In addition it addresses how these interchanges should be designed in order to ensure that target groups, i.e. the aged, youth, physically and mentally handicapped people and 'time poor' women can adequately benefit from these interchanges.

The project is being undertaken by an international consortium led by the Universidad Politecnica de Madrid (UPM) in Spain.

As part of this project we are interviewing stakeholders involved in interchanges to better understand what the key design and operational features of a successful interchange are that could be introduced elsewhere. This questionnaire is designed to help supplement information being gathered from other sources and to develop guidelines for wider use.

We understand that you may not be able to complete all of the questions and that quantitative data may be in a different format, or use different definitions. Please just answer the questions the best you can with information you have available.

PRACTITIONER INTERVIEW FORM

Date of interview:

Name of interviewer:

Organisation of interviewer:

Name of interviewee:

Type of practitioner (delete as appropriate): Transport Provider/Local Transport Authority/Business Community

Position in and name of organisation:

Country:

Question 1: Which interchange are you describing? (Please state its name and briefly describe its location and its surrounding area.)

Question 2: When was the interchange opened? When was it last refurbished/redeveloped? Please provide a brief explanation of the changes made.

Question 3: Which modes does the interchange cater for? (Please tick all that apply.)
 □ Walking
 □ Cycling (with cycle parking)
 □ Cycle hire
 □ Motorcycles/scooters/mopeds
 □ Buses
 □ Long-distance coaches
 □ Metro
 □ Light rail/ tram
 □ Heavy rail
 □ Private cars (with car parking)

☐ Private cars (with drop-off)
☐ Taxis
☐ Other

If other, please provide details.

Question 4: Please describe the interchange's role/place in the overall transport network. For example, is the interchange for local, regional, national or international connections etc.

Question 5: Please can you provide some information on current passenger numbers? Including the total passengers by mode, the percentage split by mode, the approximate share of transfers between modes and spatial scale (please see table below) and the distribution between men and women travellers.

Approximate share of transfer between spatial scales: Percentage of passengers.

NB. Spatial scales may be different for each interchange, please therefore outline the basis used for your categories e.g. local is up to 10 km, regional is up to 50 km, etc., or local covers all metro travel, regional covers travel on X bus route, national covers all X rail route.

Modes From\To	Local	Regional	National	International
Local				
Regional				
National				
International				

Please provide us with information on how you have defined the different spatial scales.

Local defined as:

[ ]

Regional defined as:

[ ]

National defined as:

[ ]

International defined as:

[ ]

Approximate share of transfer between modes: Percentage of passengers

Modes From/To	Train	Metro	Tram	Bus: Local Regional Interurban	Car	Cycling and walking	Other (specify)	Sum
Train								
Metro								
Tram								
Bus: • Local • Regional • Interurban								
Car								
Cycling and walking								
Other (specify)								
Sum								

Question 6: How many rail routes, bus routes, metro lines and tramway lines use the interchange?

Rail routes =

Bus routes =

Metro lines =

Tramway lines =

Question 7: What are the average frequencies (in minutes) for public transport arriving and departing at the terminal? What are the average frequencies (in minutes) for public transport arriving and departing during rush hour?

Rail =

Bus =

Metro =

Tram =

Question 8: Please describe the connectivity provided between modes at the terminal, i.e. in terms of ease of transfer between modes and distance between modes (in metres).

Question 9: Are there regular delays with public transport that cause difficulties with transferring between modes (i.e. once a month or so)?

Question 10: Please describe the regulatory framework within which the interchange operates. For example, who regulates the different modes and/or the interchange itself? Please also describe any regulatory barriers to interchange between modes.

Question 11: What is the ownership structure of the interchange?
   □ Public
   □ Private
   □ Joint venture (Public and Private)
   □ Other (Please provide details)

Please describe the ownership structure. If a joint venture, please provide details of the proportion that is publicly/privately owned.

```
┌──────────────────────────────────────────────┐
│                                                │
│                                                │
└──────────────────────────────────────────────┘
```

Question 12: Which organisation(s) is/are responsible for the management of the interchange? (Please tick all who are involved and describe below.)
☐ Central government or one of its agencies
☐ Regional government or one of its agencies
☐ Local transport authority
☐ Rail/Metro/LRT operator
☐ Bus operator
☐ Other (Please provide details)

Please describe the management structure.

```
┌──────────────────────────────────────────────┐
│                                                │
│                                                │
└──────────────────────────────────────────────┘
```

Question 13: Is there cooperation between the different operators for ensuring connectivity between modes? For example, relating to time-tabling, ticketing or information etc. If yes, please describe. If no, is there a reason why not?

```
┌──────────────────────────────────────────────┐
│                                                │
│                                                │
└──────────────────────────────────────────────┘
```

Question 14: Are there any factors that could facilitate cooperation between modes?

```
┌──────────────────────────────────────────────┐
│                                                │
│                                                │
└──────────────────────────────────────────────┘
```

Question 15: Who bears the financial responsibility of the interchange (maintenance, investments, local charges)?
☐ Public
☐ Private
☐ Joint venture (Public and Private)
☐ Other (Please provide details)

Please describe the responsibilities…

```
┌──────────────────────────────────────────────┐
│                                                │
│                                                │
└──────────────────────────────────────────────┘
```

Question 16: Is the interchange financially profitable?
- ☐ Yes
- ☐ No
- ☐ Don't know

Question 17: Are there any (financial) reports available?
- ☐ Yes
- ☐ No
- ☐ Don't know

Question 18: Which organisation(s) was/were responsible for the design of the multimodal interchange (including any re-developments)? (Please tick all who are involved and describe below.)
- ☐ Central government or one of its agencies
- ☐ Regional government or one of its agencies
- ☐ Local transport authority
- ☐ Rail/Metro/LRT operator
- ☐ Bus operator
- ☐ Private organisation
- ☐ Other (Please provide details)

Please describe who was responsible for design and any re-developments which have subsequently occurred...

```



```

Question 19: Please now consider the capital costs involved with developing the interchange.

  a. What was the financing model used to fund the development of the interchange?

```



```

  b. What was the expected payback time of the investment?

```



```

  c. Who were the main financiers?

```



```

Question 20: Is there a business model* developed for the interchange?

*By business model we mean a strategic plan that outlines how to utilise the interchange facilities in a way that optimises the potential revenue from existing facilities and provides cost-effective maintenance services.
☐ Yes
☐ No

If yes, please could you provide a copy. (This will be treated as confidential.)
If not, please describe how decisions on pricing and level of services are determined?

```



```

If no business model exists, do you think the interchange would benefit from having one?
☐ Yes
☐ No

Question 21: Was the public consulted during the design and/or redevelopment of the interchange?
☐ Yes
☐ No

If yes, please describe the process applied to involve the public...

```



```

Question 22: Is the public involved in any ongoing engagement with regards to the operation of the interchange?
☐ Yes
☐ No

If yes, please describe the process applied to involve the public...

```



```

Question 23: Why is this interchange considered successful? (Please tick all that apply and describe below.)
☐ Ownership/management structure
☐ Availability of interchange space
☐ Quality of transfer between modes

☐ Quality of waiting areas
☐ Range of retail establishments
☐ Security and safety
☐ Facilities for the mobility impaired
☐ Quality of journey planning and real time information
☐ Integrated ticketing arrangements
☐ Other (Please provide details)

Please describe...

```

```

Question 24: What were the key factors that influenced the design/redesign of the interchange?

1. ....
2. ....
3. ....

Question 25: Can you describe any specific methods that were used or provide guidelines that aided the coordination between modes at the multimodal interchange?

```

```

Question 26: Was energy efficiency considered in the interchange design and in its operation?
☐ Yes
☐ No

If yes, please explain how energy efficiency was ensured in the interchange design and its operation (e.g. energy use in the terminal).

```

```

If yes, regarding operation, please explain how the energy use and carbon footprint (or CO_2 emissions) of the interchange is monitored. If applicable, is it possible to estimate the percentage of alternative energies used?

```

```

Question 27: Does the interchange analyse its impact on air quality? Is air pollution considered a problem for travellers at the interchange? Has the interchange implemented any measures to improve air quality (e.g. monitoring, ventilation systems, instructions to switch off engines while waiting)? Please describe.

```
┌──────────────────────────────────────────────────────────────┐
│                                                                │
│                                                                │
└──────────────────────────────────────────────────────────────┘
```

Question 28: Are you satisfied with the information and intelligent systems in the interchange?

```
┌──────────────────────────────────────────────────────────────┐
│                                                                │
│                                                                │
└──────────────────────────────────────────────────────────────┘
```

If not, how would you improve the quality, content or provided systems and services? Please tick a) the ones currently in use and b) what you think would be essential to implement.

In use	Needed	Intelligent system or service in the interchange area
☐	☐	Journey planner for local public transport for pre-trip planning
☐	☐	Journey planner for long-distance public transport for pre-trip planning
☐	☐	Information for interchange facilities and layout available on the internet (or via call centre) for pre-trip planning (important especially for the disabled)
☐	☐	Smart ticketing (speeds up transfer)
☐	☐	Electronic departure time displays based on *timetables* (for multiple stops)
☐	☐	Electronic departure time displays based on *timetables* (at stops)
☐	☐	Electronic departure time displays based on *real-time information* (for multiple stops, incl. fleet monitoring systems)
☐	☐	Electronic departure time displays based on *real-time information* (at stops)
☐	☐	Departure times via audio calls
☐	☐	Real-time disturbance information provided via *displays*
☐	☐	Real-time disturbance information provided via *audio calls*
☐	☐	Multi-language information
☐	☐	Public access information kiosk/internet kiosk restricted for Public Transport information (not for open internet surfing)
☐	☐	Information centre with personal service

☐	☐	Audio services for the visually impaired (e.g. a special dedicated information area with a push button)
☐	☐	Guidance and warning surfaces for the visually impaired
☐	☐	Tactile maps of the interchange for the visually impaired
☐	☐	Information with hearing aids (e.g. "T-coil")
☐	☐	Matrix bar codes (e.g. QR-codes) for additional information with mobile phones (e.g. for departure times for a specific stop or platform)
☐	☐	Intelligent Indoor-Navigation System
☐	☐	Intelligent security systems (e.g. CCTV)
☐	☐	Area or terminal fleet management with the aid of cameras, in-vehicle systems, Variable Message Signs etc. for guiding buses, taxis, park&ride etc.
☐	☐	Intelligent automated passenger or people counting (infrared, video, thermal etc.)

Please provide any additional comments.

Question 29: What impact has the interchange had on employment?

a. Please describe any direct employment effects (i.e. staff needed to operate and maintain the interchange).

b. Please describe any indirect employment effects (i.e. supporting services created in the interchange).

c. Please describe an impact on the surrounding areas (i.e. new services generated in the proximity of the interchange [estimate, if no data available]).

Question 30: If possible please provide an estimate of the typical cost of housing and retail units at the interchange, and in close proximity to the interchange.

[ ]

Question 31: Have there been any changes in the amount of new start-up businesses close to the interchange?

[ ]

Question 32: Have there been any changes connected to housing in close vicinity to the interchange?

[ ]

Question 33: Has any new housing been developed in/or near to the interchange? If possible please provide the area (in m²) and the type of housing.

[ ]

Question 34: Please give an indication of the area (in m²) of commercial centres or retail in/or near to the interchange.

[ ]

Question 35: Is there good access provided between the interchange and the commercial/retail centre?

[ ]

Question 36: Have any new offices been developed in/or near to the interchange? If possible please provide the area (in m²) and the type of offices (e.g. headquarters, international or national offices).

[ ]

Question 37: Can you estimate the number or percentage of these new housing, retail or office developments which are directly as a result of the interchange?

Question 38: Can you provide any other examples of successful multi-modal interchanges in your country?
  ☐ Yes
  ☐ No

If yes, please provide details of the location, a brief description (e.g. modes of transport available) and explain in what ways the interchange is successful; any specific factors, e.g. information systems, accessibility, energy efficient design/operation.

Question 39: Please describe any particular challenges that are commonly faced in the design of multimodal interchanges?

Many thanks for completing this questionnaire. If you would like any further details on the City-HUB project please visit our website at http://www.cityhub-project.eu/.

The City-HUB Partners.

Appendix III: Travellers' attitudinal survey

This ad hoc survey was carried out within the City-HUB project in order to identify best practices in themes such as connectivity between modes, information systems and distribution of facilities and services.

Its objective was to capture the views, preferences and level of satisfaction of travellers regarding different features of an interchange. This helped to verify the main drivers of intermodal travel behaviour and identify current levels of satisfaction with the current interchange services.

The survey was conducted at the five selected pilot case studies with a collected data sample of over 2,000 travellers in total. Primarily, the population inside the interchange was characterised for each case study by age and gender. In order to obtain a valid sample of the population, a minimum number by group was collected so the survey was defined by stratified random sampling. Some questions were customised for each case study, as highlighted in the questionnaire.

The survey form reproduced here includes three parts:

Part A Travellers' satisfaction questionnaire
Part B Your trip
Part C Socioeconomic information

Please write the SURVEY NUMBER as it appears in your card: ☐

Part A Travellers' Satisfaction Questionnaire
  1. Your satisfaction level with regard to TRAVEL INFORMATION:

	– Level of satisfaction +				
	1	2	3	4	5
Availability and ease of use of travel information (timetables, routes…) at the interchange	☐	☐	☐	☐	☐

	1	2	3	4	5
Availability of travel information (timetables, routes…) before your trip	☐	☐	☐	☐	☐
Accuracy and reliability of travel information displays for bus/trains/ metro at the interchange	☐	☐	☐	☐	☐
Ticket purchase (ticket offices, automatic machines, etc.)	☐	☐	☐	☐	☐

2. Your satisfaction level with information on how to find your way around the station and associated transport facilities:

	1	2	3	4	5
Signposting to different facilities and services (retail, catering facilities, waiting areas, toilets, etc.)	☐	☐	☐	☐	☐
Signposting to transfer between transport modes in all parts of the interchange. E.g. to buses, metro, taxis, cycle parking, etc.	☐	☐	☐	☐	☐
Information and assistance provided by staff, e.g. at customer information desks	☐	☐	☐	☐	☐

3. Your satisfaction level with regard to TIME & MOVEMENT aspects inside the interchange:

	1	2	3	4	5
Transfer distances between different transport modes. E.g. to buses, metro, taxis, cycle parking, etc.	☐	☐	☐	☐	☐
Coordination between different transport operators or transport services	☐	☐	☐	☐	☐
Use of your time (transferring & waiting) at the interchange	☐	☐	☐	☐	☐
Distance between the facilities and services (retail, catering facilities, waiting areas, toilets, etc.)	☐	☐	☐	☐	☐
Number of elevators, escalators and moving walkways	☐	☐	☐	☐	☐
Ease of movement due to number of people inside the interchange	☐	☐	☐	☐	☐

4. Your satisfaction level with regard to ACCESS:

	1	2	3	4	5
Ease of access to the interchange	☐	☐	☐	☐	☐

5. Your satisfaction level with regard to COMFORT & CONVENIENCE:

	1	2	3	4	5
General cleanliness of the interchange	☐	☐	☐	☐	☐
Temperature, shelter from rain and wind, ventilation, air conditioning	☐	☐	☐	☐	☐
General level of noise of the interchange	☐	☐	☐	☐	☐
Air quality, pollution. E.g. emissions from vehicles	☐	☐	☐	☐	☐
Number and variety of shops	☐	☐	☐	☐	☐
Number and variety of coffee-shops and restaurants	☐	☐	☐	☐	☐
Availability of cash machines	☐	☐	☐	☐	☐
Availability of seating	☐	☐	☐	☐	☐
Availability of telephone signal and Wi-Fi	☐	☐	☐	☐	☐

6. Your satisfaction level with regard to the attractiveness of the station and associated transport facilities:

	1	2	3	4	5
The surrounding area is pleasant	☐	☐	☐	☐	☐
The internal design of the interchange (visual appearance, attractiveness, etc.)	☐	☐	☐	☐	☐
The external design of the interchange (visual appearance, attractiveness, etc.)	☐	☐	☐	☐	☐

7. Your satisfaction level with regard to SAFETY & SECURITY:

	1	2	3	4	5
You feel safe getting on and off the transport mode (train, bus, metro, etc.)	☐	☐	☐	☐	☐
You feel safe while inside the interchange	☐	☐	☐	☐	☐

You feel secure in the transfer & waiting areas (during the day)	☐	☐	☐	☐	☐
You feel secure in the transfer & waiting areas (during the evening/night)	☐	☐	☐	☐	☐
Lighting	☐	☐	☐	☐	☐

8. Please, give a final overall value of your satisfaction with the service of this interchange:

(Strongly dissatisfied) 1	2	3	4	(Strongly satisfied) 5
☐	☐	☐	☐	☐

Part B Your Trip

9. When you were invited to participate in this questionnaire, were you? TICK ONE ONLY

☐	Starting your journey
☐	Transferring
☐	Ending your journey

10. What was the main purpose of your journey today? TICK ONE ONLY

☐	Work (trip to work and working trip)
☐	Education
☐	Leisure or visiting family and friends
☐	Other

11. Did you travel...?

☐	Alone
☐	With a pram/pushchair
☐	With children
☐	With luggage
☐	With senior people
☐	Accompanying another adult requiring assistance because of impaired vision or mobility

12. Do you have any kind of disability?

☐	No	☐	Temporary	☐	Permanent
		Specify:			

13. Which was the overall duration of your trip (from origin to destination)?

---

14. How did you travel TO this interchange (please state your previous transport mode)?
    (Customized for each pilot case study)

☐	Long distance train	☐	Taxi
☐	Commuter train	☐	Car as driver
☐	Tramway	☐	Car as passenger
☐	Long distance coach	☐	Motorcycle/Moped
☐	Local bus	☐	Bicycle
☐	Subway	☐	Walking
☐	Airport bus	☐	Other, please name:

15. How long was your trip TO this interchange (from the origin)?

TOTAL TIME (minutes)

16. How are you travelling FROM this interchange (please state your next transport mode)?
    (Customized for each pilot case study)

☐	Long distance train	☐	Taxi
☐	Commuter train	☐	Car (driver)
☐	Tramway	☐	Car (passenger)
☐	Long distance coach	☐	Motorcycle/Moped
☐	Local bus	☐	Bicycle
☐	Subway	☐	Walking
☐	Airport bus	☐	Other, please name:

17. How long will your trip FROM this interchange (to your final destination) be?

TOTAL TIME (minutes)

18. Number of total transfers in your trip: [   ]

19. Public transport ticket used for your trip:

☐	Single ticket
☐	Daily/Weekly/monthly/yearly pass
☐	(Customised for each case study)
☐	Other (specify)

20. How much time did you spend within the interchange?

TOTAL TIME (minutes)

How did this time break down? Please provide your answer in minutes.

TRANSFERING MODES*	Queuing for boarding	Shopping	OTHER ACTIVITIES**

*TRANSFERING MODES: Walking between different transport modes for boarding. E.g. transferring between train and bus stop.
**OTHER ACTIVITIES: Buying tickets, seated at café, seated at waiting areas, etc.

21. How often do you use this interchange? TICK ONE ONLY

☐	Daily (more than 4 days)
☐	3 or 4 times a week
☐	Once or twice a week
☐	Few times a month
☐	Less frequently

Part C Socioeconomic Information

22. Gender:

☐	Male	☐	Female

23. Do you have...?

☐	A driving license
☐	Access to own car
☐	Access to a motorcycle/moped
☐	Access to a bicycle

24. How old are you?

☐	17 or less
☐	18 to 25
☐	26 to 40
☐	41 to 65
☐	66 or more
☐	Would prefer not say

25. Education level achieved:
    (Customized for each case study)

☐	Primary school
☐	High school
☐	University degree

26. What is your employment status?

☐	Employed
☐	Unemployed
☐	Retired
☐	Student
☐	Housewife/-man
☐	Other (specify)

27. Household Net Income per month
(Customized for each case study based on the minimum revenue of each country)

☐	≤1,300€* (≤2 times minimum revenue)
☐	1,300–2,600€* (2–4 times minimum revenue)
☐	≥2,600€* (≥4 times minimum revenue)

(minimum revenue for Spain: 650€)

Thank you very much for your kind cooperation.

Finally we would like you to ask your opinion on WHICH YOU THINK ARE THE 3 MOST IMPORTANT ASPECTS OF AN INTERCHANGE:

	Tick only 3
Information: trip and interchange	
Waiting Areas	
Safety & Security	
Services (toilets, ticket purchase, luggage store, etc.)	
Shops and Cafes	
Transfer communication between transport modes	
Access to the interchange	
Other(Please specify): _____	

The results of the Prize Draw will appear in the City-HUB webpage (www.cityhub-project.eu) in 2 months' time.

If you want to receive information of the results directly, please write your contact email address:

E-mail: _____

THANK YOU FOR YOUR TIME

Appendix IV: Examples of business models

CASE STUDY EXAMPLE 1: INTERCITY COACHES OF MAGNESIA INTERCHANGE IN THE SUBURB OF VOLOS, GREECE

Interchange size

Table A4.1 Dimension A assessment – Magnesia interchange

Dimension A aspects	Levels	Need for space in the interchange	Score levels	Case study score
Demand (users/ day)	<30,000	Low	1	1
	30–120,000	Medium	2	
	>120,000	High	3	
Modes of transport	Dominant – bus	Low	1	1
	Dominant – rail	Medium	2	
	Several modes and lines	High	3	
Services and facilities	Kiosks, vending machines	Low	1	1
	Several shops and basic facilities	Medium	2	
	Integrated shopping mall with all facilities	High	3	
			Total	3

→ Interchange size: small

Interchange typology

Table A4.2 Dimension B assessment – Magnesia interchange

Dimension B aspects	Levels	Upgrading level	Value	Case study value
Location in the city	Suburbs	Less	–	–
	City access	Neutral	o	
	City centre	More	+	
Surrounding area features	Non-supporting activities	Less	–	–
	Supporting activities	Neutral	o	
	Strongly supporting activities	More	+	
Development plan	None	Less	–	–
	Existing	Neutral	o	
	Existing and including intermodality in the area	More	+	
			Total	3–

→ Interchange typology: cold/hot

Business model

	Offer side		Demand side		
Key stakeholders	Services	Value propositions	Interaction with users		Users' characteristics
1 Transport operator for: • Long-distance trips • Suburban services	Services for 9 big cities (destinations out of Magnesia), 36 destinations in Magnesia, Routes from and to the airport of Volos 'Nea Anchialos National Airport'.	Governance: one transport operator so no governance issue for managing the Magnesia Bus interchange. Some cafés are located nearby the Magnesia bus interchange.	The access between the interchange and the commercial/retail centre is considered quite good, since people may use the local buses, or they can walk or cycle along the quays of the city's port. Although the interchange is very close to the university campuses, they are well segregated by a creek, so there is no interaction between the traffic generated by the campuses and the respective traffic generated by the interchange. Online tools for pre-trip planning and mobile apps.		2,703 passengers/day: • 559 passengers travelling to Athens • 406 passengers to Thessaloniki • 525 passengers to other Greek cities (i.e. Larisa, Patra, Kozani, Ioannina, Agrinio, Lamia, Trikala) • 1,213 trips have their destinations in Magnesia
	Resources		Atmosphere		
	Free Wi-Fi access. Some cafés.		No sheltered waiting areas. No passengers segregated from moving traffic. Schedules and location information on displays and incident information on loudspeakers.		
	Costs			Revenues	
No detailed information is available. One single operator pays for the cost of maintenance of the interchange.			Transport operators' fees. Commercial rents (cafes, vending machines, etc.).		

Figure A4.1 Business model scheme – Magnesia interchange.

CASE STUDY EXAMPLE II: MONCLOA, MADRID (SPAIN)

Interchange size

Table A4.3 Dimension A assessment – Moncloa interchange

Dimension A aspects	Levels	Need for space in the interchange	Score levels	Case study score
Demand (users/day)	<30,000	Low	1	
	30–120,000	Medium	2	
	>120,000	High	3	3
Modes of transport	Dominant – bus	Low	1	
	Dominant – rail	Medium	2	
	Several modes and lines	High	3	3
Services and facilities	Kiosks, vending machines	Low	1	
	Several shops and basic facilities	Medium	2	2
	Integrated shopping mall with all facilities	High	3	
			Total	8

→ Interchange size: city landmark

Interchange typology

Table A4.4 Dimension B assessment – Moncloa interchange

Dimension B aspects	Levels	Upgrading level	Value	Case study value
Location in the city	Suburbs	Less	–	
	City access	Neutral	o	o
	City centre	More	+	
Surrounding area features	Non-supporting activities	Less	–	
	Supporting activities	Neutral	o	o
	Strongly supporting activities	More	+	
Development plan	None	Less	–	
	Existing	Neutral	o	
	Existing and including intermodality in the area	More	+	+
			Total	1+

→ Interchange typology: partially integrated

Business model

	Offer side		Demand side	
Key stakeholders	*Services*	*Value propositions*	*Interaction with users*	*Users' characteristics*
12 transport operators: • Metro of Madrid: Regional authority ownership and integrated in Madrid Regional Transport Authority (MRTA). • Madrid Municipal Transport Company: Local authority ownership, also integrated in MRTA. • 9 Metropolitan bus operators. • 1 long-distance bus operator. Retail and businesses.	Metro (2 lines). Local bus (20 lines). Metropolitan bus (56 lines). Long-distance bus (2 lines). Integrated ticketing. Shopping and food services. *Resources* Four levels (3 underground): Different platforms for each mode. Waiting areas physically separated from bus operating zone. Several spacious entrances with numerous ticketing machines and personal assistance as well. Public Transport Authority (PTA) assistance office. Free Wi-Fi access. Shops and cafes.	Private concessionaire for interchange management. Governance through competition. Quality retail standards. Coordinated management with other interchanges. Madrid Interchanges Plan (existing interchange expansion). Seamless transfer among modes. Easy, legible and consistent way-finding and information. Comfortable and safe waiting areas. Dedicated bus access to the interchange from the highway.	City entrance, good access to the city centre. Listed buildings, highway access and green park in the surrounding area. Shops and restaurants nearby. Online tools for pre-trip planning and mobile apps. *Atmosphere* Well lit and maintained. Sheltered waiting areas. Passengers segregated from moving bus traffic. Access for all. Clear and integrated way-finding. Colour identification strategy. Schedules and location information on displays and incident information on loudspeakers. No integrated modal information on screens.	287,000 passengers/day: • 134,000 metro • 100,000 metropolitan bus • 53,000 urban bus Balanced by gender. Young-adult demand. Not high income level profile. Habitual users for work/study commuting.
Costs			*Revenues*	
Investment: initial construction and refurbishment. Last Moncloa refurbishment: 112.78 million euro. Maintenance costs: maintenance staff, equipment, cleaning, electricity, and so on. Management costs: management personnel, monitoring and assessment tools, and so on. Operating costs: security, information, ticketing and control staff and equipment, and so on.			Transport operators' fees. Commercial rents (shops and cafes, vending machines, etc.) Advertising fees. Madrid PTA compensates when the expected minimum demand, according to the concession agreement, is not reached.	

Figure A4.2 Business model scheme – Moncloa interchange.

CASE STUDY EXAMPLE III: KAMPPI, HELSINKI (FINLAND)

Interchange size

Table A4.5 Dimension A assessment – Kamppi interchange

Dimension A aspects	Levels	Need for space in the interchange	Score	Case study score
Demand (users/day)	<30,000	Low	1	
	30–120,000	Medium	2	2
	>120,000	High	3	
Modes of transport	Dominant – bus	Low	1	
	Dominant – rail	Medium	2	
	Several modes and lines	High	3	3
Services and facilities	Kiosks, vending machines	Low	1	
	Several shops and basic facilities	Medium	2	
	Integrated shopping mall with all facilities	High	3	3
			Total	8

→ Interchange size: city landmark

Interchange typology

Table A4.6 Dimension B assessment – Kamppi interchange

Dimension B aspects	Levels	Upgrading level	Value	Case study value
Location in the city	Suburbs	Less	−	
	City access	Neutral	o	
	City centre	More	+	+
Surrounding area features	Non-supporting activities	Less	−	
	Supporting activities	Neutral	o	
	Strongly supporting activities	More	+	+
Development plan	None	Less	−	
	Existing	Neutral	o	
	Existing and including intermodality in the area	More	+	+
			Total	3+

→ Interchange typology: fully integrated

Business model

	Offer side		Demand side	
Key stakeholders	*Services*	*Value propositions*	*Interaction with users*	*Users' characteristics*
Transport operators at Kamppi building: • Metro • Western bus lines • Tram Commercial centre Rental apartments and offices	21 local bus lines. 40 regional bus lines. 15 metropolitan bus lines. 1 international bus line. 1 metro line. 2 tram lines. Bicycle rental.	PPP for construction of the building and managing commercial area. Underground bus access to the interchange from the western road network, avoiding congested access to the city. Good integration with commercial activities. Pedestrian access from 3 different streets. Need to improve connections and information about services located at central rail station.	Located in the city centre, adjacent to the main commercial area. Short stay at interchange: 42% less than 5 min.	64% female users Adults: 50% >40 years 79% medium-high levels income Trip purpose: 55% work, 30% leisure Medium-long distance bus trips All type of frequency users
	Resources		*Atmosphere*	
	Downtown location. Taxi stops. 750 parking spots underground. Bicycle Centre: offers parking and maintenance. 500 m underground walking tunnel connection with Central Railway Station, where North and East bus lines depart. Big commercial centre with retail shops and restaurant.		Well lit and maintained. Waiting areas are located in the commercial area next to gates. Passengers are segregated from platforms by sliding doors. Access for all. Clear and integrated way-finding. Visible enforcement of security, with video surveillance, police and even a prison-room. Schedules on displays and gates. Not integrated modes information on screens. Automatic ticketing machines and NFC payment systems.	
Costs			*Revenues*	
Maintenance costs: maintenance staff, equipment, cleaning, electricity, and so on. Management costs: management personal, monitoring and assessment tools, and so on. Operating costs: security, information, ticketing and control staff and equipment, and so on.			Transport operators' fees. Commercial rents (shops, cafes and restaurants, vending machines, etc.). Advertisements rents. Apartments and offices for rent.	

Figure A4.3 Business model scheme – Kamppi interchange.

Appendix V: Workshops and stakeholders

The involvement of different stakeholders in the City-HUB project, such as local and regional authorities, transport operators and end-user organisations, included the organisation of some events. Three workshops were held throughout the project lifetime following the focus group technique in order to interchange ideas and find a balance between the interests of different stakeholders. A final conference took place at the end of the project which presented its outcomes and also focused on reaching potential users of the results and the transferability of the findings.

Workshop and final conference details are listed in Table A5.1 and are followed by the stakeholders who participated in them, apart from the members of the consortium work team (Table A5.2).

Table A5.1 List of City-HUB workshops

Workshop	Date	City
1st City-HUB Workshop	21 March 2013	Budapest (Hungary)
2nd City-HUB Workshop	3 February 2014	London (UK)
3rd City-HUB Workshop	9 October 2014	Thessaloniki (Greece)
City-HUB Final Conference	19 February 2015	Lille (France)

Table A5.2 List of stakeholders participating in City-HUB workshops

Name	Affiliation	Country
Abreu, Joao	IST-Lisbon	Portugal
Aggelakakis, Aggelos	Centre for Research and Technology Hellas (CERTH)/Hellenic Institute of Transport (HIT)	Greece
Ágó, Mátyás	Municipality of Érd	Hungary
		(Continued)

Table A5.2 (Continued) List of stakeholders participating in City-HUB workshops

Name	Affiliation	Country
Aldecoa, Javier	Consorcio Regional de Transportes de Madrid	Spain
Alegre Valls, Lluís	Metropolitan Transport Authority of Barcelona	Spain
Baggs, Jonathan	Crossrail	United Kingdom
Balabekou, Ifigeneia	OASTH (Thessaloniki Urban Transport Organisation)	Greece
Barta, Eszter	Miskolc Városi Közlekedési Zrt. (MVK)	Hungary
Bennett, Simon	Crossrail	United Kingdom
Bessmann, Erik	Institut Français des Sciences et Technologies des Transports, de l'Aménagement et des Réseaux (IFSTTAR)	France
Boudi, Zakaryae	Institut Français des Sciences et Technologies des Transports, de l'Aménagement et des Réseaux (IFSTTAR)	France
Bourbotte, Daniel	Institut Français des Sciences et Technologies des Transports, de l'Aménagement et des Réseaux (IFSTTAR)	France
Chatzigeorgiou, Anna	Volos Municipality	Greece
Chatzigeorgiou, Christos	OASTH (Thessaloniki Urban Transport Organisation)	Greece
Chatzilamprou, Ismini	TRAINOSE S.A.	Greece
Chmela, Petr	ROPID (Regional Organiser of Prague Integrated Transport)	Czech Republic
Chrisohoou, Evi	Centre for Research and Technology Hellas (CERTH)/Hellenic Institute of Transport (HIT)	Greece
Chrisostomou, Katerina	Centre for Research and Technology Hellas (CERTH)/Hellenic Institute of Transport (HIT)	Greece
Cré, Ivo	POLIS	International
Crest, Thierry du	Lille Metropole Urban Community	France
Cserni, Gabriella	Miskolc Holding Zrt.	Hungary
de Oña, Juan	Universidad Politécnica de Granada	Spain
Devecz, Miklós	Miskolc City Transportation Company (MVK)	Hungary
Fejes, Balázs	Centre for Budapest Transport (BKK)	Hungary
Földesi, Erzsébet	MEOSZ – National Federation of Disabled Persons' Associations	Hungary
Gaitanidou, Lila	Centre for Research and Technology Hellas (CERTH)/Hellenic Institute of Transport (HIT)	Greece

Table A5.2 (Continued) List of stakeholders participating in City-HUB workshops

Name	Affiliation	Country
Gebhardt, Laura	German Aerospace Centre Institute of Transport Research	Germany
Giannakakis, Konstantinos	TAXI Thessalonikis ERMIS	Greece
Gogas, Michael	Centre for Research and Technology Hellas (CERTH)/Hellenic Institute of Transport (HIT)	Greece
González Álvarez, Antonio	Agence d"Urbanisme Bordeaux métropole Aquitaine	France
Granquist, Tom	Akershus County Council	Norway
Haralambidou, Sofia	University of Thessaly	Greece
Hashizumé, Anne	RATP (Régie Autonome des Transports Parisiens)	France
Hasiak, Sophie	Centre for studies and expertise on Risks, Environment, Mobility, and Urban and Country Planning (CEREMA)	France
Horner, Donald	Network Rail	United Kingdom
Ioannidou, Anna-Maria	Centre for Research and Technology Hellas (CERTH)/Hellenic Institute of Transport (HIT)	Greece
Iordanopoulos, Panagiotis	Centre for Research and Technology Hellas (CERTH)/Hellenic Institute of Transport (HIT)	Greece
Jansen op de Haar, Constantijn	Stadsregio Amsterdam	the Netherlands
Jhamb, Divij	Transport Research Laboratory Limited (TRL)	United Kingdom
Karagiannis, Konstantinos	Municipality of Volos, Greece	Greece
Kormányos, László	MÁV-START Zrt. (Hungarian National Railways)	Hungary
Korsós, Mónika	Municipality of Érd	Hungary
Kotoula, Nilia	Centre for Research and Technology Hellas (CERTH)/Hellenic Institute of Transport (HIT)	Greece
Laurent, Gilles	Fédération Nationale des Usagers des Transports (FNAUT)	France
Lázár, József	Centre for Budapest Transport (BKK)	Hungary
Lychounas, Michalis	Kavala Municipality	Greece
Mitsakis, Evangelos	Centre for Research and Technology Hellas (CERTH)/Hellenic Institute of Transport (HIT)	Greece
Myrovali, Glikeria	Centre for Research and Technology Hellas (CERTH)/Hellenic Institute of Transport (HIT)	Greece

(Continued)

Table A5.2 (Continued) List of stakeholders participating in City-HUB workshops

Name	Affiliation	Country
Nagy, Istvan	Miskolc Holding Zrt.	Hungary
Norbert, Merkel	Miskolc Holding Zrt.	Hungary
Oszter, Vilmos	KTI Közlekedéstudományi Intézet Nonprofit Kft (KTI)	Hungary
Papadopoulos, Symeon	OASTH (Thessaloniki Urban Transport Organisation)	Greece
Papaioannou, Panagiotis	Aristotle University of Thessaloniki	Greece
Papaix, Claire	Institut Français des Sciences et Technologies des Transports, de l'Aménagement et des Réseaux (IFSTTAR)	France
Papoutsis, Konstantinos	Centre for Research and Technology Hellas (CERTH)/Hellenic Institute of Transport (HIT)	Greece
Perger, Imre	MÁV-START Zrt. (Hungarian National Railways)	Hungary
Ploeg, Ruud van der	European Metropolitan Transport Authorities (EMTA)	International
Politis, Ioannis	Aristotle University of Thessaloniki	Greece
Rabuel, Sébastien	Nantes Métropole	France
Rigaud, Philippe	Institut Français des Sciences et Technologies des Transports, de l'Aménagement et des Réseaux (IFSTTAR)	France
Salanova, Josep Maria	Centre for Research and Technology Hellas (CERTH)/Hellenic Institute of Transport (HIT)	Greece
Schellaert, Valérie	Institut Français des Sciences et Technologies des Transports, de l'Aménagement et des Réseaux (IFSTTAR)	France
Sheldon, Alan	Network Rail	United Kingdom
Simandl, Aleš	Dopravní podnik hl. m. Prahy, a.s.	Czech Republic
Smith, James	Transport for London (TfL)	United Kingdom
Spanos, Georgios	OASTH (Thessaloniki Urban Transport Organisation)	Greece
Stathakopoulos, Alexander	Centre for Research and Technology Hellas (CERTH)/Hellenic Institute of Transport (HIT)	Greece
Tendli, Krisztina	TRENECON COWI Consulting and Planning Ltd.	Hungary
Toskas, Ioannis	ATTIKO METRO S.A.	Greece
Tromaras, Alkiviadis	Centre for Research and Technology Hellas (CERTH)/Hellenic Institute of Transport (HIT)	Greece

Table A5.2 (Continued) List of stakeholders participating in City-HUB workshops

Name	Affiliation	Country
Valóczi, Dénes	Budapest University of Technology and Economics (BME)	Hungary
van Keulen, Heleen	Regio Utrecht	the Netherlands
Vanhanen, Kerkko	HSL Helsinki Region Transport	Finland
Vernooij, Mette	Stadsregio Amsterdam	the Netherlands
Wright, Ian	Passenger Focus	United Kingdom
Zarras, Konstantinos	Municipality of Xanthi, Greece	Greece

Index